Francisco Javier Gonzalez Martinez

Performance of new GNSS satellite clocks

Performance of new GNSS satellite clocks

by
Francisco Javier Gonzalez Martinez

Dissertation, Karlsruher Institut für Technologie (KIT)
Fakultät für Bauingenieur-, Geo- und Umweltwissenschaften, 2013
Referenten: Prof. Dr.-Ing. Bernhard Heck, Prof. Dr. Dr. phil. nat. Urs Hugentobler

Impressum

Scientific
Publishing

Karlsruher Institut für Technologie (KIT)
KIT Scientific Publishing
Straße am Forum 2
D-76131 Karlsruhe

KIT Scientific Publishing is a registered trademark of Karlsruhe
Institute of Technology. Reprint using the book cover is not allowed.

www.ksp.kit.edu

Print on Demand 2014

ISBN 978-3-7315-0112-1

Performance of new GNSS satellite clocks

Zur Erlangung des akademischen Grades eines

DOKTOR-INGENIEURS

DOKTORS DER NATURWISSENSCHAFTEN

von der Fakultät für

Bauingenieur-, Geo- und Umweltwissenschaften
der Universität Fridericiana zu Karlsruhe (TH)

genehmigte

DISSERTATION

von

Dipl.-Ing. Francisco Javier Gonzalez Martinez

aus Villanueva de la Reina, Spanien

Tag der mündlichen Prüfung: 07 Juni 2013

Hauptreferent: Prof. Dr. -Ing. Bernhard Heck

Korreferent: Prof. Dr. Dr.phil.nat.Urs Hugentobler

Karlsruhe   2013

To my family and especially to my wife, Giuliana.

Acknowledgements

First and foremost, I would like to express my deepest gratitude to my advisor Prof. Bernard Heck for his brilliant supervision and his continual guidance and support throughout my Ph.D. years. Without his encouragement during the initial difficult steps to start this thesis and his technical insight, advice and continuous support, this dissertation and my further career development would have never been possible. Secondly, I want to express my gratitude to Prof. Urs Hugentobler for his deep technical review of this manuscript. It has been a real pleasure and privilege for me to have both as advisors.

Secondly, I would like to acknowledge the European Space Agency (ESA) for the access to Galileo In Orbit Validation Element (GIOVE) data. It has been an encouraging and stimulating opportunity to be at ESA during the launch and test of GIOVE satellites. The in depth review of the core elements of a test Global Navigation Satellite System (GNSS), such as GIOVE-mission, has been a unique opportunity for me to understand full scale systems. Most likely the new clocks, signals and modulations tested in GIOVE and now available in Galileo will change the face of GNSS in the near future.

Special mention should go to all my GIOVE-Mission colleagues for their support in numerous areas of the thesis, such as: atomic clock technology (Pierre Waller and Joerg Hahn); limit of time transfer (Patrizia Tavella, Ilaria Sesia and Giancarlo Cerretto); receiver operation (Massimo Crisci and Gustavo Lopez); about Precise Orbit Determination (Daniel Navarro) or about end-to-end operations (Richard, Andreas and Gaetano). Special thanks also to Beatriz Moreno, it was really fun to develop a positioning software together from scratch and review all the error contributions until arriving at centimetre accuracy.

Finally, I want to express my deepest gratitude to my parents for their support in my decision to leave my job in Spain and become a student again looking for knowledge in Europe. Special thanks go to my dearest sister Eva and her lovely family for their unconditional love. A reminder goes also to my grandfather who was always there for his emigrated grandson. I am especially indebted to my wife Giuliana for her love, her patience and boundless encouragements despite all the time stolen from her during the course of this thesis. I look forward to hearing her reading ancient Greek histories under the Mediterranean sun during our next holidays...

Zusammenfassung

Die Uhren an Bord von Satelliten globaler Navigationssatellitensysteme (GNSS) sind zentrale Elemente, aus denen Zeit- und Navigationssignale erzeugt werden. Die Leistung der Navigationssysteme hängt unter anderem von der Leistung der Uhren sowie von der Fähigkeit des Systems ab, das Verhalten der Uhren einzuschätzen und vorherzusagen. Diese Bedeutung wurde bereits am US-amerikanischen Global Positioning System (GPS) erkannt. Dieses nutzte von deutschen Bodensystemen abgeleitete Technik für die ersten weltraumqualifizierten Rubidiumuhren in Block IIA Satelliten. Diese wurden unter Beibehaltung redundanter Verfügbarkeit mit Cäsiumtechnik in den neueren IIR und -IIF Blöcken konsolidiert.

Der Begriff 'Uhr' wird in der Regel für den Frequenzstandard an Bord des Satelliten selbst dann verwendet, wenn sie keine direkte Zeitinformation liefert. Das Frequenzsignal wird weiter von der Elektronik modifiziert, bevor es vom Satelliten ausgesendet wird. Nur das Navigationssignal schließt echte Zeitinformation ein. Diese Arbeit klärt, dass der Ausdruck 'Zeitsignal' am Ausgang der Navigationsantenne angemessener ist. Das mit der Zeitinformation modulierte Zeitsignal wird vom Empfänger in Phasen- und Codemessungen zurückgewonnen. Dieses Konzept ermöglicht die Unterscheidung zwischen dem Atomuhr-Frequenzstandard 'physische Uhr', dem Signal am Satellitenantennenausgang für jede Frequenz 'Signaluhr' und der aus POD (Precise Orbit Determination) abgeleiteten 'ionosphärenfreien oder scheinbaren Uhr'.

Heutzutage sind Zeitsignale überall. Sowohl bei zeitspezifischer Laborausstattung als auch bei Massenmarktanwendungen wird die Zeitübertragung zwischen Punkten durch Einweg- oder Zwei-Wege-Techniken durchgeführt. GNSS-Zeitübertragung ist ein klares Beispiel von Einweg-Zeitübertragung. Die Quelle (A) sendet ein Zeitsignal an den Benutzer (B) über ein Übertragungsmedium mit einer Verzögerung (d) über einen Übertragungsweg. Die Korrektur der Wegverzögerung erfordert die Berechnung der Positionen von A und B sowie der Laufzeitverzögerungen über den Weg mit hoher Genauigkeit.
GNSS-Systeme berechnen die Position und Rückverfolgbarkeit zwischen der Satellitenzeit (A) und der Systemzeit durch geodätische Zeitübertragungstechniken, um den Zeitsignaloffset an den Benutzer (B) zu liefern. Geodätische Zeitübertragung ist auch das genaueste Mittel im Messwesen, um Zeit und Frequenz zwischen entfernten Zeitslabors, welche für die Erzeugung nationaler oder internationaler Zeitreferenzen wie UTC verantwortlich sind, zu übertragen. Vor der Analyse der Uhrleistung in der Umlaufbahn untersucht diese Dissertation die Methodik und

Genauigkeit, die mittels geodätischer Zeitübertragung erreicht wird, um Grenzen und mögliche Verbesserungen zu identifizieren.

Die Überprüfung der Methodik zeigt, dass bei der Berechnung der Satellitenposition die Zeitschätzung stark von der Umlaufbahn abhängt. Für einen typischen Empfänger hoher Genauigkeit mit Rundstrahlantenne liegt das erwartete theoretische Limit der Einwegzeitübertragung bei Nutzung der Codeinformation bei 100ps (1σ), und 1 ps (1σ) bei Auswertung der Phaseninformation. Der Gewichtsfaktor für die Zeitschätzung ist von diesem 1/100-Verhältnis abgeleitet. Dieser Faktor bewirkt, dass der Absolutwert der Zeitübertragung von den Code- und die Genauigkeit von den Phaseninformationen abhängt. In der Praxis sind mit dem besten Stand der Technik geodätische Zeitübertragungen mit 70 ps (rms) und 20 ps (1σ) möglich, wahrend die Frequenzübertragung liegt zwischen 1E-12 $\tau^{-1/2}$ (vor $\tau = 1$ Sekunde) bis 1E-15 (vor $\tau = 10^{6}$).

GPS ist nicht das einzige Navigationssystem. GLONASS hat die volle orbitale Konstellation von 24 Satelliten im Jahr 2011 wiederhergestellt, und das Galileo-System ist im Aufbau. Der erste Start eines Galileo-Satelliten erfolgte im Oktober 2011, der zweite im Oktober 2012. Diese dienten der Systemvalidierung. Die volle Konstellation wird ab 2013 sukzessive aufgebaut. Während die Galileo-Konstellation im Aufbau begriffen ist, hat die GIOVE-Mission schon ab 2005 im Weltraum demonstriert, wie Galileo funktionieren wird. GIOVE-Satelliten tragen eine neuartige Rubidiumatomuhr (RAFS) und den ersten passiven Wasserstoffmaser (PHM) an Bord eines Navigationssatelliten, welche eine gute Extrapolation der endgültigen Uhrleistung in der Umlaufbahn erlauben.

Bereits vor dem Start von GIOVE-B war klar, dass die Leistung des PHM an die Grenzen des Stands der Technik von geodätischen Zeitübertragungsmöglichkeiten und darüber hinaus den Fähigkeiten des Bodensegments stoßen würde. Aufgrund der begrenzten Anzahl der Stationen sowie der Instabilität von Hardwareverzögerungen erreicht die geodätische Zeitübertragungsleistung der GIOVE-Mission 0.5 ns (rms), 0.3 ns (1σ) sowie 2.2E-12 $\tau^{-1/2}$ die beste Stabilität. Dieser Wert ist zweimal schlechter als die erwarteten 1E-12$\tau^{-1/2}$ für das PHM und liegt auf dem Niveau der besten RAFS in GIOVE-Satelliten. Vorrangiges Ziel der GIOVE-Mission war die Sicherung der von der internationalen Telekommunikationsbehörde vergebenen Frequenzen. Ein weiteres Ziel war die Bestätigung der in Galileo zu fliegenden Nutzlastausrüstung. Insbesondere da zuvor keine europäische Atomuhr weltraumerprobt war, wurde ihre Validierung das anschließende Hauptziel der GIOVE-Mission. Eine andere Methode war erforderlich, um die Leistung der GIOVE-Uhren im Orbit zu überprüfen.

In dieser Arbeit wird eine neuartige Methode vorgeschlagen, beschrieben, in Software implementiert und mit einer ausgezeichneten Übereinstimmung validiert anhand von GPS-Satelliten durch einen Vergleich mit öffentlich zugänglichen IGS-Ergebnissen. Das kurzfristige Verhalten unter 300 Sekunden wird nicht durch IGS-Endprodukte abgedeckt. Die Kombination dieser Methode mit POD-Ergebnissen hat außerdem die volle Charakterisierung von GNSS-Uhren zum ersten Mal erlaubt. Die bestätigte Eignung für die Charakterisierung von GNSS-Uhren gestattete die Anwendung auf GIOVE-Uhren. Es wurde gezeigt, wie die kurzfristige Stabilität der RAFS und PHM im Einklang mit den Bodenmessungen sind. Dabei war es sogar möglich, die aktivierten RAFS-Einheiten aus der Messung zu identifizieren. Diese gute Übereinstimmung hat die Validierung dieser neuartigen Methode sowie die erste volle Charakterisierung von GNSS-Uhren und die erfolgreiche Erreichung des zweiten Ziels der GIOVE-Mission ermöglicht. Diese neue Methodik wurde durch andere Gruppen wie CNES [39] oder DLR [112] verwendet und angepasst, mit vergleichbaren Ergebnissen.

Die einzige unbekannte im PHM beobachtete Wirkung war eine harmonische Komponente mit 0.5 ns Amplitude in der geschätzten ionosphärenfreien Uhr. Während harmonische Komponenten in GPS-Uhren ein bekanntes Merkmal sind [158], erwähnt nur eine neue Veröffentlichung die Temperatur kurz als den Ursprung dieser Wirkung [150], lässt aber eine tiefergehende Analyse vermissen. Der Ursprung der Harmonischen in den scheinbaren Uhren von GNSS-Satelliten wird in dieser Arbeit überprüft und geklärt. Es wird gezeigt, dass die Amplitude für die meisten Satelliten mit dem Winkel der Sonne bezüglich der Bahnebene in Beziehung steht. Diese Korrelation zeigt eine mögliche Abhängigkeit von der Temperatur an. Es wird eine einfache Methode vorgeschlagen, welche die erwartete harmonische Welle von der Empfindlichkeit der physischen Uhren in Bezug auf die Temperatur ableitet. Die gute Übereinstimmung zwischen erwarteten und beobachteten Werten zeigt, dass Harmonische in der scheinbaren Uhr von GNSS-Satelliten hauptsächlich durch die thermische Empfindlichkeit der physchen Uhr verursacht werden.

Die einzige Unstimmigkeit besteht beim PHM, wo temperaturinduzierte Schwankungen im Atomfrequenzstandard in Anbetracht von zahlreichen Hinweisen unwahrscheinlich erscheinen. Die harmonische Komponente wurde schon vor dem Satellitenstart als Resultat eines künstlichen Effekts aufgrund der mit nur 13 vorgesehenen Bodenstationen erzielbaren Bahnmessgenauigkeit vermutet. Diese Hypothese wurde später bestätigt durch eine verminderte Amplitude bei Erhöhung der Anzahl von Messungen durch Hinzufügen von Stationen, SLR Messungen oder durch die Verlängerung der Bogenlänge der Satellitenbahnen.

Die Umlaufperiode deutet auf das Strahlungsdruck-Modell (SRP: Solar Radiation Pressure) als wahrscheinliche Ursache der Harmonischen in der PHM-Schätzung hin. Es kann sein, dass das für die SRP-Schätzung verwendete empirische Modell aufgrund der niedrigen Anzahl von Stationen ungenau oder von der ungünstigen Geometrie betroffen ist. Sobald die Galileo-Konstellation vollständig ist und eine höhere Anzahl von Sensorstationen verfügbar wird, sollte die Genauigkeit des SRP-Modells für Galileo-Satelliten im PHM-Modus überprüft werden.

Die besondere Aufmerksamkeit, die hier der harmonischen Welle zuteil wird, ist nicht trivial. Deren Auswirkung betrifft die dem Nutzer gelieferte Uhrprädiktion. Diese stellt immer noch einen der größeren die Fehlerbeiträge für Echtzeitnavigation und die Hauptbeschränkung für längeren Ephemeridengebrauch dar. Uhrkorrekturen sind daher auch der wichtigste Mehrwert der Echtzeitdienste. Die harmonische Komponente soll in Richtung der Einführung von möglichen Reduktionsstrategien auf System- oder Benutzerebene verstanden werden.

Unabhängig vom Ursprung der harmonischen Welle wird hier gezeigt, wie die Einbeziehung von harmonischen Koeffizienten in der Uhrvorhersage die Genauigkeit steigert und ein sinnvolles stochastisches Modell liefert. Eine Verbesserung wird hauptsächlich in den Polynomtermen beobachtet, was ermöglicht, auf die Übermittlung der harmonischen Koeffizienten an den Benutzer zu verzichten. Im PHM-Modus ist der Vorhersagefehler auf dem gleichen Niveau wie das Schätzungsrauschen (0.3 ns, 1σ) bei 100 min und auf der Ebene der harmonischen Komponente (0.5 ns) bei einem Tag. Auf wissenschaftlicher Seite hat die überlegene Frequenzwiederholbarkeit der vom PHM gelieferten neuen Uhrtechnik erlaubt, die erwartete relativistische Frequenzänderung (4.718E-10) mit einem Fehler von 1.2% zu messen (5.58E-12). Außerdem hat die gegenwärtig angewandte periodische relativistische Korrektur einen periodischen Fehler von 0.1 ns, wie in [83] festgestellt. Diese bei anderen GNSS-Uhren verdeckte Wirkung ist mit dem PHM eindeutig sichtbar. Diese Tatsache demonstriert, wie der neue PHM den Uhrfehler unterhalb anderer Fehlerquellen gebracht hat.

Während sich GNSS-Konstellationen langsam weiterentwickeln, wird die neue Generation von optischen Uhren am Boden entwickelt und verspricht eine um mehrere Größenordnungen bessere Leistung (bis auf 1E-18 Ebene). Es bleibt die Frage zu beantworten, welche neuen Möglichkeiten diese verbesserten Uhren liefern. Im Prinzip öffnet es ihre Verwendung für Satelliten oder Referenzstationen die Möglichkeit, ein funktionales Modell für die Uhrenschätzung und ggf. -prädiktion zu verwenden. Die Uhrparameter stellen 80-90% der Unbekannten dar. Die Reduktion auf drei Parameter reduziert die Anzahl von Unbekannten drastisch, und die Korrelation mit anderen geschätzten Parametern verringert demzufolge auch die Notwendigkeit einer großen Anzahl von Stationen um Satellitenprodukte zu berechnen. PHMs in Galileo-Satelliten und H-Maser auf Bodenstationen bestätigten diese Annahme bereits. Dennoch wurde die Instabilität von Gruppenverzögerungen als potentiell störender Einfluss identifiziert, der noch stets sorgfältig in der Berechnung berücksichtigt werden muss. Ein zusätzlicher Gewinn wird nicht in der Schätzung, sondern in der Vorhersage und Zeitkontrolle erwartet. Wenn sich die Uhrstabilität nur um eine Grössenordnung verbessert, wäre es bei der Vorhersage möglich, den Uhrfehlerbeitrag in Navigationssystemem vollständig zu eliminieren und Echtzeitdienste unnötig zu machen. Für die Zeitkontrolle erübrigte sich die Notwendigkeit, die Boden- und Satellitenuhren zu steuern. Letztendlich würden die größten Vorteile in der Erzeugung der Systemzeit zum Tragen kommen, die gegenwärtig komplexe Uhrenensembles am Boden erfordert.

Während optische Uhren mittelfristig nicht für GNSS-Systeme zu erwarten sind, erzeugen sie ein großes Interesse im Bereich der Forschung, und es wird empfohlen, diese Technologie in eine der fundamentalen Physikmissionen der ESA [49] einfließen zu lassen, um einer zukünftigen Verwendung in GNSS den Weg zu bereiten.

Acronyms

Below the acronyms used in this dissertation are presented. They are also defined when they first appear in the text.

1PPS	One Pulse per Second
AC	Analysis Center
ADEV	Allan Deviation
AFS	Atomic Frequency Standard
AIUB	Astronomical Institute of the University of Bern
BGD	Broadcast Group Delay
BIPM	Bureau International des Poids et Mesures
CDF	Cumulative Distribution Function
CDMA	Code Division Multiple Access
CMCU	Clock Monitoring and Control Unit
CNAV	Civil Navigation
CONGO	Cooperative Network for GIOVE Observation
Cs	Cesium
DDS	Direct Digital Synthesizer
DLL	Delay-Locked Loop
ECEF	Earth Centered, Earth Fixed
EQM	Engineering Qualification Model
ESA	European Space Agency
ESTEC	European Space Research and Technology Centre
ET	Ephemeris Time
FDMA	Frequency Division Multiple Access
FDU	Frequency Distribution Unit
FEI	Frequency Electronics, Inc
FGUU	Frequency Generation and Up-conversion Unit
FM	Flight Model
FOC	Full Operational Capability
FSDU	Frequency Synthesizer and Distribution Unit
FTS	Frequency and Time Systems
GD	Group Delay

GESS	Galileo Experimental Sensor Station
GETR	Galileo Experimental Test Receiver
GFZ	GeoForschungsZentrum
GGSP	Galileo Geodetic Service Provider
GIOVE	Galileo In Orbit Validation Element
GIOVE-M	GIOVE-Mission
GLONASS	Global Navigation Satellite System
GNSS	Global Navigation Satellite System
GPS	Global Positioning System
GST	Galileo System Time
GSTB	Galileo System Test Bed
GTRF	Galileo Terrestrial Reference Frame
IAU	International Astronomical Union
ICAO	International Civil Aviation Organization
ICD	Interface Control Document
IERS	International Earth Rotation and Reference Systems Service
IFB	Inter-Frequency Bias
IGS	International GNSS Service
IOV	In-Orbit Validation
ISB	Inter System Bias
MASER	Microwave Amplification by Stimulated Emission of Radiation
NAVSTAR	Navigation Satellite Timing and Ranging
NIST	National Institute of Standards and Technology
OCXO	Oven Controlled Crystal Oscillator
ODTS	Orbit Determination and Time Synchronization
OWCP	One Way Carrier Phase
PCV	Phase Center Variation
PDF	Probability Density Function
PHM	Passive Hydrogen Maser
PLL	Phase-Lock-Loop
POD	Precise Orbit Determination
PPP	Precise Point Positioning
PRN	Pseudo-random Noise
PVT	Positioning Velocity and Time
QZSS	JAXA Quasi-Zenith Satellite System
RAFS	Rubidium Atomic Frequency Standard
Rb	Rubidium

RIRT	Russian Institute of Radionavigation and Time
SBAS	Satellite-Based Augmentation System
SI	System of Units
SIS	Signal in Space
SIS-ICD	Signal in Space Interface Control Document
SLR	Satellite Laser Ranging
SN	Serial Number
SpT	Spectra Time Switzerland
SRP	Solar Radiation Pressure
SVN	Space Vehicle Number
TAI	International Atomic Time (from french: Temps Atomique International)
TCXO	Temperature Controlled Crystal Oscillator
TKS	Time Keeping System
TM	Telemetry
TOR	Time of Reception
TOT	Time of Transmission
TOW	Time of Week
TSP	Time Service Provider
TT	Terrestrial Time
TTFF	Time To First Fix
TWSTFT	Two-Way Satellite Time and Frequency Transfer
UERE	User Equivalent Range Error
URE	User Range Error
US	United States
USA	United States of America
USNO	United States Naval Observatory
UTC	Universal Time Coordinated
VCXO	Voltage-Controlled Crystal Oscillator
VLBI	Very Long Baseline Interferometry
WN	Week Number
XO	Crystal Oscillator

Contents

1 Introduction

In Global Navigation Satellite Systems (GNSS), the on-board clocks are a key component from which timing and navigation signals are generated. The performance of the navigation systems rely on, amongst other factors, the performance of the clocks, as well as the capability of the system to estimate and predict the clock behaviour.

This importance was recognized at an early stage by the leader system, the US Global Positioning System (GPS) which first adapted ground technology for the first space-qualified rubidium clocks in Block-IIA (from German technology), then further consolidated in the latest Block-IIR and -IIF while keeping also dual source availability with cesium technology. Even if atomic frequency standard technology has improved steadily over the last 30 years, this technology is currently only mature enough and space qualified by a limited number of suppliers in some countries: cesium standards are available in Russia, USA and Europe; rubidium standards in USA and Europe; and passive hydrogen standards only in Europe. All GNSS systems, even if not directly under military control, have clear applications for the military domain which restrict the exportability of the technology between countries. Other global or regional systems, such as the Chinese COMPASS or Japanese Quasi-Zenith Satellite System (QZSS) rely currently on this foreign technology but are currently being developed in order to establish their own atomic frequency standards.

The first Galileo launch took place on 21 October 2011 with the first Russian Soyuz ever launched from French Guyana. This first launch marked the start of the deployment of the full constellation. Payload on-board these Galileo satellites has been fully tested on the Galileo In Orbit Validation Elements GIOVE-A and -B launched in 2005 and 2008 respectively. These validation satellites were fully representative in terms of payload equipment. In particular, they carried a new type of Rubidium Atomic Frequency Standard (RAFS) and the first Passive Hydrogen Maser (PHM) on-board a navigation satellite, allowing a good extrapolation of Galileo performance in orbit. Since the launch of GIOVE-A and GIOVE-B, over 7 years of cumulated in-orbit operations have confirmed the maturity of the new atomic clock technologies and paved the way for operational Galileo satellites. Before the GIOVE-B launch, it was already clear that the performance of the new PHM clock would be above the state-of-the-art geodetic time transfer capabilities, as later confirmed once in-orbit. Performance of on-board clocks has been reported in numerous publications for limited periods, sometimes mixed with other topics. However, a complete and continuous overview of these satellites' clock performance is missing for a complete understanding of the new possibilities provided by these new clocks. This thesis provides this missing overall view of GIOVE satellite clocks and proposes a new methodology to verify the PHM performance in orbit.

On the GPS side, the first GPS Block IIF satellite was launched on 28 May 2010 carrying a new enhanced RAFS, a new cesium clock and a new timing subsystem. Some of the new clocks aboard Galileo and GPS promise to bring the clock error contribution below other error sources such as the orbit error. This hypothesis needs to be tested and the state of the art in geodetic time transfer reviewed to identify other potential new opportunities offered by these new clocks.

Satellite navigation is now a reality and is part of our daily lives. Most of the devices that are available today in the market are single frequency for mass-market, and double frequency for precise users. New devices have started to be equipped with multiple GNSS interfaces. Thus, it is reasonable to assume that after a second and third frequency become available, the new GNSS devices will also be equipped with multiple frequency radio-navigation interfaces. The new mass-market services will use double frequency, and precise users will become multi-frequency based. Envisaging the new complexity associated with the new frequencies, modulations and systems, one of the central questions treated in this thesis is a clarification of the relationship between the different 'clock concepts' currently being used.

Additionally, the main subject of this thesis is the new 'clocks' and the relation to the new modulation 'signals' available in new GNSS satellites: Galileo, GPS Block-IIF, GLONASS-K and COMPASS.

Outline of the dissertation

Chapter 2 - GIOVE mission

In this chapter, the GIOVE mission is briefly presented. Further details about GIOVE mission objectives and elements can be found in numerous related publications available on the GIOVE website (www.giove.esa.int). Complementary independent networks and estimations are also briefly reviewed as they represent a valuable complement to the reference products obtained in the core of GIOVE mission.

Chapter 3 - Time scales involved in GNSS

GNSS systems can be seen considered in the context of time transfer between different time scales. Traceability between UTC, system and satellite time is computed on-ground and broadcast to the user:

$$UTC(k) \rightarrow T_{SYST} \rightarrow t^{sat} \rightarrow t_{rec}$$

Time transfer performed between clocks moving in different reference frames is affected by relativistic effects related to the invariance of the speed of light. Each time an atomic clock is activated after some non-operational period, some deviations with respect to the predicted initial nominal frequency is expected. The new PHM clock technology on-board the Galileo spacecraft provides, after switch-on, an unprecedented initial level of frequency accuracy, particularly in comparison with previous technologies, allowing an accurate measurement of the net relativistic frequency shift.

Chapter 4 - Timing signals realization

The timing signal broadcast by a GNSS satellite is not only derived from the atomic frequency standards. This chapter intends to produce a more complete understanding of the satellite timing subsystem by examining the physical components, history, new and future trends of its components. This understanding is an absolute necessity in order to explore the possibilities offered by the new atomic frequency standards, signals and modulations on board the new GNSS satellites.

Chapter 5 - Methodology applied in geodetic time transfer

In this chapter the GNSS time transfer methodology between the time scales is investigated. The understanding of the time transfer helps to reveal some choices implemented in the physical realization of the timing signals and to estimate the theoretical accuracy limit of the geodetic time transfer.

Firstly, methods are briefly reviewed together with IGS product combinations normally used as benchmark. Secondly, the ionosphere free combination is studied and the other parameters included in the estimated 'ionosphere-free clock' identified. From these parameters the group delay is identified as the main bias. Thirdly, the estimation of group delays together with ionosphere estimations is analysed. Finally, a practical example of inter-frequency bias and inter-system bias estimation is also presented with GIOVE satellites with standard and novel methodologies.

Chapter 6 - Performance of Geodetic time transfer

This chapter analyses the precision and accuracy of GNSS time transfer by reviewing the quality of the clock estimations performed in the GIOVE mission for GPS and Galileo satellites. The methodology and results for GPS are cross-checked with IGS. An analysis is performed step-by-step internally from measurement residuals and repeatability, and externally against different software packages, different data networks, reference measurements and independent techniques such as Satellite Laser Ranging (SLR) or Two-Way Satellite Time and Frequency Transfer (TWSTFT). From this analysis it is possible to observe how the traditional methodology for clock stability assessment is limited in terms of short term coverage and noise floor. An innovative methodology is proposed to complement the assessment on this area.

Chapter 7 - Harmonics in satellite clocks

All GNSS signal clocks show periodic fluctuations long realized since the first estimates of GPS clocks. Nevertheless, their characteristics have been only recently characterized without any clear identification of their origin. In this chapter the origin of GNSS clock harmonics is analysed and clarified.

Chapter 8 - GNSS clock stability and prediction

This chapter reviews the overall GNSS clock performance and prediction capabilities. Clocks are characterized from Precise Orbit Determination (POD) estimates in terms of stability and robustness. Finally, current clock prediction and integrity methodologies are reviewed and applied to all GNSS satellites clocks.

Chapter 9 - Conclusions and outlook

Finally the last chapter collects the main conclusions on the current status, new findings and novel methodologies proposed in the dissertation. To conclude, the future new opportunities brought by the new clocks are summarized.

2 GIOVE Mission

2.1 Introduction

GNSS systems analysed in this dissertation are GPS, GLONASS and Galileo. The COMPASS system will be addressed only whenever reliable information exists due to the lack of transparency and the absence of a public interface document. As Galileo will be considered the demonstration satellites GIOVE-A and -B, carrying the same atomic clocks envisaged in the final operational payload. At the time of launch of these satellites in 2005 no commercial receiver or permanent station was able to track the new signals. A ground network of 13 globally distributed stations was established hosting a flexible experimental receiver developed by Septentrio able to track the new signals and modulations. Observations are regularly collected at the GIOVE processing center in The Netherlands where the navigation message is also generated.

Initially no other globally distributed network was able to track GIOVE satellites. Network data from the GIOVE mission were first provided in 2008 to the Galileo Geodetic Service Provider (GGSP), a scientific consortium in charge of generating geodetic products for Galileo resulting in several weeks being processed using scientific standards. Then, the complete month of December 2009 was made freely available to the overall scientific community resulting in several publications. Commercial receivers were finally available to the public after the signal in space Interface Control Document (ICD) was made freely accessible in 2008. A consortium of scientific institutions joined their efforts to create a flexible separate network able to track GIOVE and other new satellites and the first complementary network appeared in 2009, becoming fully operational in 2010.

In this chapter the GIOVE mission will be briefly presented. Further details about GIOVE mission objectives and elements can be found in numerous related publications, such as the one with the initial results [136] or with the final summary [50], available at www.giove.esa.int. Complementary independent networks and estimations are also briefly reviewed as they represent a valuable complement to the reference products obtained in the core of the GIOVE mission.

2.2 Mission

The Galileo Positioning System is a satellite navigation system, being built by the European Union (EU) as an alternative to GPS and GLONASS.

As a risk mitigation activity,in 2002 the European Space Agency (ESA) started to develop an experimental ground mission segment, called Galileo System Test Bed Version 1 (GSTB-V1). Within the GSTB-V1 project, tests of Galileo orbit determination, integrity and time synchronization algorithms were conducted in order to generate navigation and integrity core products based on GPS data. In 2003, the second stage of the overall Galileo system test bed implementation began with the development of two GIOVE satellites and an associated ground segment infrastructure. The GIOVE Mission or Galileo System Test Bed Version 2 (GSTB-V2) is an experimental infrastructure for the testing of Galileo critical technologies.

The main objectives of these two satellites are to secure the use of the frequencies allocated to the Galileo system, to verify the most critical technologies of the operational Galileo system, including the on-board atomic clocks and the navigation signal generators, to characterize the novel features of the Galileo signal design (including the verification of user receivers and their resistance to interference and multipath), both on space and ground segments. In particular, the main goals are to:

- Secure the use of the frequencies allocated by the International Telecommunications Union for the Galileo system;

- Characterize the orbits to be used by the in-orbit validation satellites;

- Characterize the on-board clock (RAFS and PHM) technology in space;

- Collect lessons learned on space segment onboard units pre-development and in-orbit operations;

- Assess the performance of the navigation service (including navigation message uplink and broadcast);

- Test the overall timeliness and operational aspects (including data collection from sensor stations), data processing, message generation and uplink.

2.3 Space segment

The GIOVE-A spacecraft launched on 28th December 2005 included most of the critical equipment of the final Galileo payload, in particular the navigation signal generation unit able to generate Galileo-representative signals (L1-interplex, E6- interplex and E5-AltBOC), as well as two RAFS, from which only one RAFS can be operative while the other was kept switched off as redundant back-up in case of failure. ESA formally ended GIOVE-A's mission at the end of June 2012, although it is still being operated without L-Band transmission by prime contractor Surrey Satellite Technology Ltd of Guildford, UK, to collect radiation data and performance results from a GPS receiver.

Fig. 2.1: GIOVE-A and -B satellites. Source: www.spaceinimages.esa.int

The GIOVE-B spacecraft was launched on 25th August 2008; its payload was very similar to the one belonging to GIOVE-A, with the same capability to transmit additional modulations on L1 carrier (CBOC and TMBOC) with enhanced multipath characteristics. In addition, it included the first Passive Hydrogen Maser (PHM) frequency standard operating in medium Earth orbit, where radiation environment was particularly severe for electronic equipment. ESA formally ended GIOVE-B's mission at the end of July 2012.

Figure 2.1 presents an artist's view of the GIOVE satellites. Ground control segments, from which both satellites are operated, are located in Guildford for GIOVE-A and in Fucino for GIOVE-B.

2.4 GESS network

The GIOVE mission core infrastructure for experimentation consists mainly of a network of Galileo Experimental Sensor Stations (GESS) distributed worldwide that acquire and collect the GIOVE satellite signals at 1Hz. Two GESS are installed at time laboratories; one GESS is installed at the time laboratory located at INRiM, Turin, connected to an active hydrogen maser, located in a controlled environment. The INRiM time reference will be used as the basis for Galileo System Time (GST) in the GIOVE mission. A second GESS is installed at the United States Naval Observatory (USNO) in order to provide a link to GPS time by common view for Galileo-GPS time offset validation. Additionally, the stations at Noordwijk (GNOR and GNO2) have been updated with a hydrogen maser as their input frequency source.

The observations collected by the GESS are sent to the ground processing center located at ESTEC. The orbit determination and time synchronization software processes pseudo-range and carrier phase measurements collected from the GESSs in order to provide GIOVE and GPS orbit and clock estimates and predictions. Predictions are further quantified and converted in the experimental navigation message to be broadcast by both GIOVE satellites. In the following sections this closed loop will be presented (see Figure 8.11).

In the GIOVE mission, a combined GPS/Galileo receiver is used. The Galileo Experimental Test Receiver (GETR) developed by Septentrio is a 54-channel dual-constellation multi-

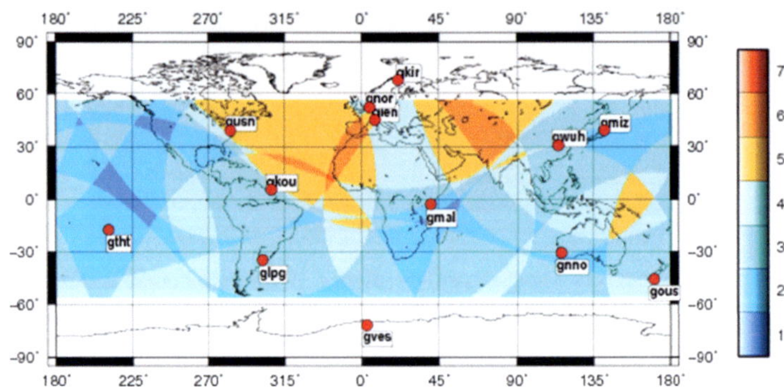

Fig. 2.2: GESS network coverage (2008) [114].

frequency receiver that is capable of tracking GPS L1, L2 and L5 and Galileo L1, E5a, E5b, E5 (AltBOC) and E6 signals, and provides detailed measurements and data for all tracked signals. It can operate in dual constellation GPS/Galileo mode as well as in Galileo-only mode. In its current version, it contains 8 Galileo channels in addition to the 48 GPS channels tracked by a separated board. The GETR is a customized receiver specially developed for ESA but shares practically the same design as the commercial version called GeNeRx1 which is available to other users [151]. A detailed overview of the design can be found in the GeNeRx datasheet [151] and several Septentrio publications [174, 173]. The availability of a ground mission segment for the GIOVE mission at the time of the receiver design and especially at the launch date was not guaranteed. In order to cope with a possible satellite time scale not synchronized with the system time, the GETR is able to work without the decoding of the navigation message; this particular feature will be later explained when looking at the receiver measurements.

Quadband Space Engineering antennas are used by the stations using GETR receivers. These antennas provide excellent performance in terms of group delay (GD) and phase center stability when operating in extreme environmental conditions. This antenna technology is also used in EGNOS and Galileo operational stations. The original GESS network was composed of 13 stations - its coverage is depicted in Figure 2.2 in terms of number of receivers observing a GIOVE satellite with respect to its projection on the Earth. This network was later expanded with 3 additional stations hosting one receiver and antenna developed by Novatel. Both items were initially tested and modified into the currently installed versions from 2010 in the network, which are also available to the general public [117, 118]. The Noordwijk site holds both types of receiver and antenna chains in common clock (GNOR and GNO2), with the possibility of antenna selection in order to perform zero baseline tests. Further information about the GESS elements can be found on its datasheet [74].

Fig. 2.3: CONGO network (2011). Source: www.weblab.dlr.de

2.5 Independent networks and software

2.5.1 CONGO network

The Cooperative Network for GIOVE Observation (CONGO) was established in 2008 by the German Space Operations Center (DLR/GSOC) and the Federal Agency for Cartography and Geodesy (BKG) as an early test bed for experimentation with new GNSS signals. The CONGO network rapidly increased in number from 8 sites available in 2010 till 19 sites in mid-2011. The distribution of sites is plotted in Figure 2.3.

The CONGO network employs three different types of multi-frequency multi-constellation receivers: the Septentrio GeNeRx1 receiver, the Javad Triumph Delta-G2T/G3TH receivers and the Leica GRX1200+GNSS receiver. The receivers are fed by different types of antennas: the Leica's AR25 chokering antenna, the Leica AX1203+GNSS survey antenna and Trimble's Zephyr Geodetic II. At the Wettzell site one Leica AR25 antenna feeds three different receivers.

CONGO network data represent an excellent complement to the GIOVE mission providing a fully independent network of stations with different commercial antennas and receivers. The processing of this network alone allows the verification of GIOVE satellite estimations. The combination of GIOVE and CONGO networks duplicates the number of stations allowing the estimation of an enhanced solution and the assessment of the impact of a higher number of stations in the orbit determination processing.

2.5.2 GGSP

The Galileo Geodetic Service Provider (GGSP) was a project fully-funded by the sixth framework programme of the European Community. The goal of the GGSP was to develop a prototype for the generation of the Galileo Terrestrial Reference Frame (GTRF) and the establishment of a service with products and information for any potential users.

The GGSP was a highly qualified consortium of European and non-European experts in the field of geodesy with the main objective being the development of the reference frame and the generation of a service with data, products and information relevant for the potential Galileo users. The project started in July 2005 and finished in May 2009 with the initial realization of an experimental reference frame realization based on GPS/Galileo measurements from the IGS and GESS stations.

All partners were experienced analysis centers of IGS with routine contributions. Each center used different independent software (Bernese, EPOS and NAPEOS) with higher flexibility than operational software used in the GIOVE mission and updated as soon as any advance had been found leading to the improvement of the products. Furthermore, different algorithms and processing strategies were applied by each center which provided an internal validation of the estimations. Final combination should lead to the best possible solution for GIOVE.

During the GGSP experimentation GPS and GIOVE satellite orbits and clocks were estimated. GIOVE estimation covered a limited period of four reference GPS weeks (1500, 1505, 1509 and 1515) processed by all analysis centers, whereas the Astronomical Institute of the University of Bern (AIUB) covered a longer continuous period (weeks 1500-1520). Under http://www.ggsp.eu/ggsp_home.html it is possible to find the geodetic products available for download, together with the reports and strategies used. Detailed information about each member can be found on the homepages of each partner.

The products were obtained in a combined adjustment based on IGS and GIOVE mission networks. In any combined solution the number of GPS observations is much higher as the ratio GPS/GIOVE is unbalanced by the number of stations ($>100/13$) and satellites (30/2). As a consequence, it has to be remarked that the solution is strongly based on GPS. This fact is particularly relevant as the intersystem bias instabilities in the receiver were considered to be mainly absorbed by the Galileo satellites and, as a consequence, the solution degraded with respect to a normal GPS satellite. Nonetheless, the GGSP products represent a reference estimation with the state-of-the-art processing used by the international geodetic community.

2.6 Conclusions

The first launch of In-Orbit Validation (IOV) satellites was performed in October 2011, the second in October 2012 and the deployment of the Full Operational Capability (FOC) is scheduled from 2013 on. While the Galileo constellation is being deployed, the GIOVE mission has already demonstrated in a real environment how Galileo will work.

Since the launch of GIOVE-A in 2005 and GIOVE-B in 2008, over 6 years of cumulated in-orbit and ground experimentation confirmed the maturity of the most critical technologies, the validity of analytical models and the ability to meet the challenging performance of the Galileo System. The GIOVE signal-in-space was almost fully representative of the Galileo System in terms of radio frequency and modulations, as well as chip rates and data rates. The GIOVE payload was also representative of the Galileo payload - all payload units being tested in both satellites. In particular, the GIOVE atomic clocks can be considered to be fully representative of Galileo clocks, with the exception of minor differences in the case of the RAFS which will later be explained in Chapter 4 dedicated to GNSS atomic clocks. As a consequence, the performance for Galileo clocks can be extrapolated from the GIOVE experience.

The scientific community was not in a position to independently track GIOVE satellites at the time of launch due to the new signals and modulations being used. The products obtained by the scientific community based on GESS and independent networks represent an excellent complementary data set used in this dissertation to compare the accuracy of geodetic time transfer used in the GIOVE mission with the state of the art of geodetic GNSS models.

3 Time scales involved in GNSS

3.1 Introduction

GNSS systems provide a positioning and a timing service as the solution to a four-dimensional problem in which the local position (x,y,z) and local time (t) of satellites and receivers are referred to a common reference frame (X,Y,Z) and reference time scale t_{SYS}. Some timing users also need traceability to Universal Time Coordinated (UTC), as a consequence the system time is traced to a specific realisation of UTC called UTC(k) where k is the selected time laboratory recognized by the BIPM as contributing to the creation of TAI.

Principally, each time scale is defined by an origin and a basic interval. The basic interval is the second as defined by BIPM and maintained by a local realization. The international Universal Time Coordinated is created by the BIPM as an ensemble of atomic clocks at different laboratories generating UTC(k). Each GNSS system time is generated by the corresponding ground segment based on an ensemble of atomic clocks. Satellite time is maintained by a single local atomic clock, while the receiver time is maintained normally by a crystal oscillator. The measurement of the basic interval with a different degree of accuracy makes the different time scales to deviate from each other.

The timing signal transferred from the satellite to the user provides the traceability to the satellite time, and the navigation message the traceability between the other time scales. The solution of the navigation problem allows the user to increase the accuracy of the time transfer between the receiver and system times. Finally, the recovery of UTC(k) information allows the user to make his local realization traceable to UTC:

$$UTC \rightarrow UTC(k) \rightarrow t_{SYS} \rightarrow t^{sat} \rightarrow t_{rec}$$

In this chapter, the GNSS time transfer between the time scales is examined from the data message perspective. All time scales need to be closely synchronized and the time efficiently transferred to the user to make the system work. As a consequence, timekeeping of ground and satellite time is carefully reviewed.

Time transfer performed between clocks moving in different reference frames is affected by relativistic effects related to the invariance of the speed of light. A review of the principal effects is required to understand the time transfer between the time scales realization. Until some decades ago, Einstein's equivalence principle was not widely accepted. Atomic clocks on board GNSS satellites have widely extended the application of the theory to everyday life.

The new PHM clock technology on-board the Galileo space craft provides an unprecedented frequency initialization accuracy, allowing for a more accurate demonstration and measurement than with previous atomic clocks. The measurement of the expected relativistic effect with the first PHM clock on-board GIOVE-B will be also provided in this chapter.

The analysis of the time transfer before the physical realization of the timing signal is not unintentional. The understanding of the data transfer will help to reveal some choices implemented in the physical realization of the satellite time scale explained in the next chapter.

3.2 Universal Time Coordinated

GNSS time scales and Universal Time Coordinated (UTC) are linked to each other being the actual realization of UTC supported by the GNSS time transfer between the timing laboratories which contribute to the creation of UTC. In order to understand the relationship between GNSS time scales and UTC, it is useful to briefly review the history of UTC.

The unit of time, the second, was formerly considered to be the fraction 1/86400 of the mean solar day. The exact definition of 'mean solar day' was left to the astronomical community. In 1958 the second was linked to the frequency of the cesium standard by measurement of the Ephemeris Time (ET) between the years 1954-1958 with respect to the natural resonance frequency of the cesium atom (v_0 = 9,192,631,770 Hz), making them agree on 1st January 1958 [100] with an accuracy of ± 20 Hz due to the uncertainty of the ephemeris second. This definition was used in 1968 to create the Atomic Time [51] and was formally adopted into the international System of Units (SI) by the Bureau International des Poids et Mesures (BIPM). From this date on, the second has no longer been defined in terms of astronomical motions.

Since 1970, the BIPM maintains the International Atomic Time (TAI) on the basis of the readings of atomic clocks operating in various time laboratories in accordance with the SI second as realized on the rotating geoid as the scale unit [24]. Currently, TAI is generated using data from about two hundred atomic clocks in over fifty national laboratories.

TAI drifts slowly away from ET based on the earth rotation as the earth rotation is slowing down. As a consequence one ET second requires more than v_0 cycles. After some attempts to use different frequency offsets, in its Recommendation 460 the International Radio Consultative Committee (precursor to the International Telecommunications Union) introduced the concept of leap second in order to keep the fundamental frequency constant maintaining TAI and ET aligned by only phase steps. The committee also decided to begin the new UTC system on 1 January 1972. In 1973, the General Assembly of the International Astronomical Union (IAU) recommended the use of UT1 with a maximum limit of [UT1-UTC]$< \pm 0.950$ seconds.

Finally, the definition of UTC was formally recognized in Resolution 5 of the 15th meeting of the General Conference on Weights and Measurements (1975), to be supported for civil time. Since then, the UTC scale has been derived from TAI by the insertion of leap seconds to ensure approximate agreement with the time derived from the rotation of the Earth. The choice of the dates and the announcement of the leap seconds falls under the responsibility of the International Earth Rotation and Reference Systems Service (IERS). Physical realizations of UTC - known as UTC(k) - are maintained in national metrology institutes or observatories contributing with their clock data to the BIPM. The establishment of UTC and the leap second generated, at the time, a long debate and much disagreement as the present discussion of the leap second removal shows. An excellent review of UTC history is provided by BIPM [103].

3.3 Astronomical time scales

The SI second is defined as the following: a second is the duration of 9 192 631 770 periods of the radiation corresponding to the transition between the two hyperfine levels of the ground state of the cesium 133 atom. For many years, the mean solar time measured from mean noon at Greenwich was the basis for civil and astronomical time; as explained in section 3.2, the definition of the second was linked to the Ephemeris Time (ET) by measurement of its value on the Earth surface during the years 1954-1956. The IAU still recognizes Greenwich Mean Solar Time as UT0 as observed at any location on the Earth, without regard for the location of the Earth's rotation axis with respect to the observing site. If the position of the pole with respect to the observing location is known, small corrections can be applied to produce a time scale, UT1, that is free of the local effects of the station's geography.

In 1967-68, TAI was created and its definition extended in 1980 as a coordinate time scale defined on a geocentric reference frame with the SI second realized on the rotating geoid as the scale unit. However, ET did not include any relativistic effects. It was necessary to link the time scale definitions to coordinate systems with origins at the center of the Earth and the center of the solar system, respectively, and are consistent with the general theory of relativity. In 1991 the $4th$ IAU resolution, defined Terrestrial Time (TT) as an evolution of ET and two additional relativistic time scales Geocentric Coordinate Time (TCG) and Barycentric Coordinate Time (TCB) were adopted, centered in the center of the mass of the Earth and our solar system respectively. The latest Geocentric and Barycentric time scales replaced Terrestrial Dynamical Time (TDT) and Barycentric Dynamical Time (TDB), which presented scale difference between their coordinate transformations. Each of the coordinate time scales TCB, TCG, TT and TDB can be related to the proper time τ of an observed provided that his trajectory in the Barycentric or Geocentric Coordinate reference system is known [73].

All time scales use the SI second as their basic interval, the only difference being where they are defined. TT is defined in the geoid and represents an ideal representation of TAI. The origin of TAI was estimated to be ahead of UT1 on its definition by 32.184 seconds and the origin of TT is defined as:

$$TT = TAI + 32.184 seconds$$

Whereas the GNSS or UTC time scales are enough for most users, astronomical users have to transfer the GNSS system time to TT and apply the required transformation to the desired geocentric or barycentric reference systems, where the celestial mechanics take place. The transformation from TT to the geocentric and barycentric coordinate time can be found in the IAU report [128] and the IERS recommendation [73].

3.4 System time

As explained in the introduction, a time scale is based on an origin and a basic interval. Each system time defines the origin with respect to UTC and mantains the basic interval on the ground by an ensemble of atomic frequency standards. The definition of each time scale is covered in the Interface Control Document (ICD) related to each system :

- GPS time (GPST) is established by the Control Segment and is referenced to UTC as maintained by the U.S. Naval Observatory UTC(USNO) zero time-point defined as midnight on the night of January 5, 1980/ morning of January 6, 1980 [115, 59].

- GLONASS time is generated on the basis of GLONASS Central Synchronizer (CS) time. The GLONASS time scale is periodically corrected by an integer number of seconds simultaneously with UTC corrections. Due to the leap second correction, there is no integer-second difference between GLONASS time and UTC (SU) realization in Moskva. However, there is a constant three-hour difference between these time scales [148].

$$t_{GLONASS} = UTC(SU) + 03hours \qquad [3.1]$$

- GST physical realization will be performed by the Precise Time Facility as an ensemble of 2 H-maser and 4 cesiums frequency standards [156]. Galileo System Time (GST) start epoch will be 00:00 UT on Sunday August 22nd 1999 (midnight between August 21st and 22nd). At the start epoch, GST will be ahead of UTC by thirteen (13) leap seconds [52]. It has to be highlighted that the GIOVE time origin is aligned with the GPS time definition [48].

GPS and Galileo time scales are continuous atomic time scales differing by the same number of leap seconds to UTC and by a constant -19 seconds offset to TAI. Information about the introduction of leap seconds is provided in the navigation message. Alternatively, GLONASS is a discontinuous atomic time scale. The approach used by GLONASS to include leap seconds implies discontinuities in the transmitting time which introduces further difficulties for the receiver manufacturer during the leap second introduction. Even if a dedicated annex of the GLONASS interface control document deals with the expected receiver operation during the

DD/MM/YYYY	UTC	TAI	GPS	GLO	GAL
01/01/1958	00:00:00	00:00:00			
01/01/1972	00:00:00	00:00:10			
06/01/1980	00:00:00	00:00:19	00:00:00		
01/07/1982	00:00:00	00:00:21	00:00:02	03:00:00	
22/08/1999	00:00:00	00:00:32	00:00:13	03:00:00	00:00:13
01/01/2006	00:00:00	00:00:33	00:00:14	03:00:00	00:00:14
01/01/2009	00:00:00	00:00:34	00:00:15	03:00:00	00:00:15

Tab. 3.1: GNSS-BIPM Time Scale Relation at 00^h UTC

leap second, some difficulties are still reported in the receiver processing during the leap second introduction [94].

To understand the relationship between the different time scales, it is useful to observe the time scales from the perspective of TAI, as visualized in Figure 3.1. Also useful is to observe their relation in Gregorian representation at their definition and after introduction of the last two leap seconds as reported in Table 3.1.

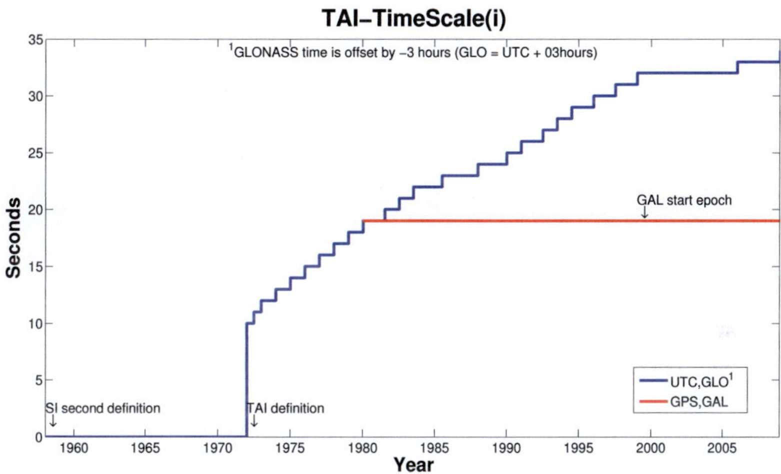

Fig. 3.1: TAI-TimeScale(i), integer offset

The navigation solution provides the difference between the receiver and system time. With the aim of supporting timing users, each system time scale is traceable to a time laboratory (k) which maintains traceability to UTC as created by the BIPM. The difference between the system time T_{SYS} and the time laboratory time scale $UTC(k)$ is provided in the navigation message through a linear model:

$$T_{UTC(k)} = T_{SYS} + A0 + A1(t - t_0)$$

[3.2]

	UTC(k)	A0 [s]			A1 [s/s]	
		max	min	1σ	max	min
GAL	TSP	2.0	9E-10	5E-9	7E-09	9E-16
GPS	USNO	2.0	9E-10	9E-8	7E-09	9E-16
GPS(L5)	USNO	9E-7	3E-11	9E-8	2E-12	4E-16
GLO	SU	1.0	8E-09	1E-3	-	-
GLO(-M)	SU	1.0	5E-10	1E-3	-	-

Tab. 3.2: $|T_{SYS}$ - UTC| information transmitted in the navigation message and the declared uncertainty

Table 3.2 provides the navigation message allocation for $T_{SYS} - UTC(k)$ correction. The minimum value represents the quantization error of the model, whereas the maximum number represents the maximum possible offset to UTC that can be corrected through the navigation message by the ground segment.

From BIPM time scale estimations (available on-line at *www.bipm.org/jsp/en/TimeFtp.jsp*) the accuracy of GPS has always been better than specified resulting in decreased limits in the latest L5 signal message definition. GLONASS-M satellites also show a slight modification by decreasing the quantization by one order of magnitude, from 8 nanoseconds to 0.5 nanoseconds resolution - more in line with GPS and Galileo definitions.

In practice, each system time is smoothly steered by delta frequency steps to UTC so that the difference between the system time and UTC remains within the lower limits specified in each signal in space ICD or in the service performance document. The offset is intended to be lower than 90 ns for GPS and 1ms for GLONASS. Galileo traceability to UTC will be performed by the Time Service Provider (TSP). The main function of this entity is to provide parameters for steering Galileo System Time (GST), as realized at the Galileo precise timing facility, with a UTC-GST time offset of less than 50 ns and uncertainty of less than 26 ns [1].

3.5 Satellite time

The satellite time transmission is linked to the following approach [115] :

1. Each satellite operates on its own local time;

2. All time-related data in the messages shall be in satellite local time;

3. All other data in the navigation message shall be relative to system time;

4. The acts of transmitting the navigation messages shall be executed by the satellite on local time.

As a consequence, local time is the time reference on each space craft to trigger all navigation related events such as the generation of spreading codes, navigation messages and time tags insertion in the navigation message. Besides the time tags all other information carried by the navigation message (ephemeris, clocks, almanac...), intended to be generated by the system, is provided in terms of system time. Several time scales are maintained inside the satellite as several elements in the payload and platform have their own internal oscillator. The concept of satellite local time used for navigation requires a clear definition.

3.5.1 Local time definition

Commercial satellites are basically composed of two parts, the platform and the payload. The platform is the general hardware in charge of maintaining the satellite attitude, thermal control, radiation shielding and payload operations through on-board and ground commands (e.g. change clocks, change signals, etc). The payload is the dedicated hardware carried on board the satellite to fulfill the mission objectives such as the frequency standard, navigation signal generation unit, etc.

Each standard satellite has a primary time on the platform maintained by the on-board computer. The payload may maintain its own time or be slaved to the platform for telemetry and telecommand operations. Platform time runs in a crystal oscillator which can drift up to several seconds per day and needs to be synchronized to the ground by space-to-ground time correlation of the telemetry and telecommand packages.

In the case of GNSS satellites, the payload carries precise atomic clocks which are in charge of the local satellite time for the navigation payload. Atomic clocks on board navigation satellites are Atomic Frequency Standards (AFS), delivering only a frequency signal. The normal procedure to distribute a timing signal in a laboratory across different hardware equipment is by means of a one pulse per second signal (1PPS) which represents a physical realization of the time scale.

The same principle is used on the satellite as depicted in the simplified Figure 3.2. The frequency standard is used as reference by a clock distribution subsystem to create the 1PPS signal. This physical signal is used to tick a time counter which can be considered as the beating clock. This same 1PPS signal is used by the signal generator unit to encode the message. Time tag information from the counter is then injected in the navigation message and transmitted to the user through the signal-in-space.

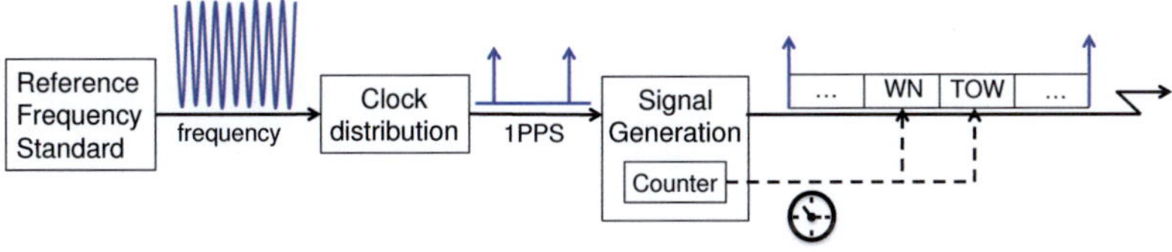

Fig. 3.2: Local satellite time generation and distribution

Operational Galileo satellites implement a similar approach provided in [29, Figure 4]. The AFS generates a 10MHz signal, which is provided to the frequency control unit for conversion to 10.23MHz and distribution. The frequency generation and the up-conversion unit up-converts the signal to the navigation frequencies (e.g. 1.5GHz) and provides the 10.23MHz signal to the navigation unit for 1PPS generation and signal modulation. The 1PPS is also provided to the on-board computer to synchronize the platform time to the payload time. GPS uses a slightly different approach by first encoding the navigation message and afterwards detecting the frame boundaries to trigger the counter (z-count), as visible in the related signal definition document [115, Figure 3-1].

In summary, the satellite local time in navigation satellites is derived from the atomic clock signal and physically created by a pulse signal used to generate the signal and to trigger a counter.

3.5.2 Time tags in the navigation message

The counter is used to time tag the messages and provide them to the user through the navigation message. This time tag information on the navigation message is required to recover the time in an absolute way and resolve the ambiguity in the pseudorange. In order to compute the pseudorange, the receiver requires the absolute time information as explained in the next section 3.6 in step 3.

In the absence of absolute time in the receiver due to a cold start or a degraded signal, the receiver starts to get ambiguous pseudoranges, the rate of repetition of time tag information will constrain the time to the first valid unambiguous pseudorange measurement. The navigation message definition, in terms of the amount of bits employed and the repetition rate of time tag information, is a key factor in reducing the time required for the user to get the time information t_{SYS} and to achieve a good time to first fix (TTFF) as analyzed in [6], where the time to first fix is decomposed in:

$$TTFF = T_{warm-up} + T_{acq} + T_{track} + T_{SYS} + T_{PVT}$$

It is nowadays clear that with modern receivers using parallel correlators the time to track T_{track} and acquire T_{acq}, the signals have strongly decreased, becoming almost negligible. Attention is turned to the time to get first system time and first valid ephemeris to compute the Positioning Velocity and Timing (PVT) solution. Receiver based techniques exist in order to achieve a navigation position with ambiguous pseudoranges, by including an additional unknown in the navigation equation [41] but requiring a precise a-priori position. Because of the T_{SYS} recovery importance, each GNSS system carefully defines the time-tag information included in the navigation message in terms of time tags size, ranges and rate repetition.

ICD	name	signal	Bits	range	unit	rate[s]
GPS	WN	NAV	10	[0,1023]	weeks	750
		CNAV	13	[0,8192]	weeks	750
	TOW	C/NAV	17	[0,100799]	6 s	6
GAL	WN	INAV	12	[0,4095]	weeks	1,20
		FNAV				10,20
	TOW	INAV	20	[0,604799]	1 s	1-10
		FNAV				10,30
GLO	N4		5	[1,31]	4-year	30
	NT		11	[0,1461]	days	30
	tk		5	[0,23]	hours	
			6	[0,59]	min	30
			1	[0,30]	sec	

Tab. 3.3: Time-tags in the navigation message

A summary of the broadcast time stamp information is provided in Table 3.3. The broadcast time stamps definition is similar in both GPS and Galileo. Both counters are divided into Week Number (WN), and Time Of the Week (TOW) inside the navigation message. The Week Number is the number of weeks from the origin of the Galileo/GPS time. As the start time of the Galileo time scale is on the first WN roll-over of GPS, the broadcast value will be the same until the next roll over of the week number (07-Apr-2019). The TOW is the time within the week in seconds. For both systems, the week starts at midnight on Saturday (24:00) to Sunday (00:00). Since GPS and Galileo are defined as having the same offset with respect to UTC, this value is exactly the same in both systems at the second level.

Despite there being similarities in respect to these previously stated factors, here the similarities end. Each system uses a different number of bits, repetition rate, and frame boundary reference. GPS uses a 29 bit counter called Z-count, enclosing WN and TOW, transmitted at a low rate every 12 minutes. A truncation to 17 bits of the TOW is provided with a higher data rate. The 17 most significant bits of TOW are transmitted at a higher data rate in the HOW word of the NAV message. It represents the local satellite TOW with 9 seconds resolution at the start of the next message subframe. To avoid the WN roll-over, three more bits are envisaged in the new L2C civil navigation (CNAV) data, thus extending the WN to a total of 13 bits [59].

Galileo uses a 32 bit counter called Galileo System Time (GST), encompassing WN and TOW in a similar way to the Z-count. Full GST is transmitted in INAV on word type 5 and spare type 0 with a repetition rate from 1 to a maximum of 20 seconds. TOW is also included as a stand-alone in word type 6 every 11 and 20 seconds. With GST, the TOW is transmitted with a minimum repetition of 1 second, and a maximum of 20 between pages 26 and 5. For FNAV with a subframe of 50 seconds divided into 5 pages the full GST is transmitted in pages 1 to 3, the TOW stand-alone in page 4, with the 5th page including the almanac without time tag.

GLONASS currently allocates only 23 bits and uses a different approach when broadcasting NT and tk counters. Nt is the calendar number of day within a four-year interval, starting from the 1st of January in a leap year. tk is the time referenced to the beginning of the frame in an HHMMSS format. This time tag definition had a four year ambiguity for a cold start of a receiver without any year information. The allocation was extended in GLONASS-M satellites with the N4 counter, with N4 being the four-year interval number starting from 1996.

The final time to recover T_{SYS} with the present ICDs definition is analyzed in [5, 6] and is provided in Table 3.4.

System	Signal	Message	t_{SYS}
Galileo	E1B	I/NAV	20.6
Galileo	E5a	F/NAV	37.5
GPS	L1 C/A	NAV	11.7
GPS	L1 C	CNAV-2	17.6
GPS	L5	CNAV	11.7

Tab. 3.4: Seconds required to read time-tags (2σ in seconds) [6].

3.5.3 Satellite to system time relation : Navigation message

The user recovers the local satellite time through the time-tag information in the navigation message. However, all other data in the navigation message is relative to system time. A relation between system and local time is therefore required:

$$t_{sys} = t^{sat} + dt_{sys}^{sat}(t_{sys})$$
[3.3]

The offset between the satellite local time and the ground system time dt_{sys}^{sat} is calculated on ground, predicted and included in the navigation message. Its behaviour is defined by three deterministic parameters and the random noise of the clock,

$$dt_{sys}^{sat}(t_{sys}) = a_0 + a_1(t_{sys} - t_{oc}) + a_2(t_{sys} - t_{oc})^2$$
[3.4]

where the three parameters of the polynomial model also have a physical meaning:

t_{oc} , is the time of clock or reference time for the clock correction expressed in system time in seconds.

t_{sys} , is the system time as maintained on ground in seconds.

t^{sat} , is the satellite time in seconds.

a_0 , is the time offset for t_{oc} in seconds.

System	bits	scale	Range	Minimum
			t_{oc} [s]	
			[days]	[s]
GLONASS	7	15	+1.3	900
GPS(NAV)	16	2^4	+11.5	16
GPS(CNAV)	11	300	+7.1	300
GALILEO	14	60	+11.3	60
GIOVE	16	2^4	+11.5	16
			a_0 [s]	
GLONASS	22	2^{-30}	$\pm$ 2.0E-3	9.3E-10
GPS	26	2^{-35}	$\pm$ 9.7E-4	2.9E-11
GALILEO	31	2^{-34}	$\pm$ 6.2E-2	5.8E-11
GIOVE	26	2^{-31}	$\pm$ 1.5E-2	4.7E-10
			a_1 [s/s]	
GLONASS	11	2^{-40}	$\pm$ 9.3E-10	9.1E-13
GPS	20	2^{-48}	$\pm$ 1.9E-09	3.6E-15
GALILEO	21	2^{-46}	$\pm$ 1.4E-08	1.4E-14
GIOVE	16	2^{-43}	$\pm$ 3.7E-09	1.1E-13
			a_2 $[s/s^2]$	
GLONASS	-	-	-	-
GPS	10	2^{-60}	$\pm$ 4.4E-16	8.7E-19
GALILEO	6	2^{-59}	$\pm$ 5.3E-17	1.7E-18
GIOVE	12	2^{-70}	$\pm$ 1.7E-18	8.4E-22

Tab. 3.5: $t^{sat} - t_{SYS}$ or clock model in each GNSS system broadcast message

a_1 , is fractional frequency offset with respect to the frequency of the system time in unit of $[s/s]$.

a_2 , is the frequency drift model of the satellite clock in unit of $[s/s^2]$.

The clock prediction associated to the model requires special attention and is analyzed later in Section 8.4. In order to broadcast the model to the user, it becomes necessary to have a "quantization" of the real number used to represent the model in Equation 3.4 as integer numbers which can be transmitted in the navigation message using the lowest possible number of bits. The declaration of bit allocations for transmission and the scale factor is provided in each signal-in-space ICD. Sign handling is the same in GPS/Galileo with two's complement encoding, with the sign bit (+ or -) occupying the most significant bit. In the case of GLONASS, the most significant bit is the sign bit. The chip "0" corresponds to the sign "+", and the chip "1" corresponds to the sign "-".

Table 3.5 provides a summary of the clock model quantization in each system. The three systems use similar approaches with different optimizations concerning the number of bits and scale. In GPS and Galileo, the definition of t_{oc} uses the same strategy with the time of clock

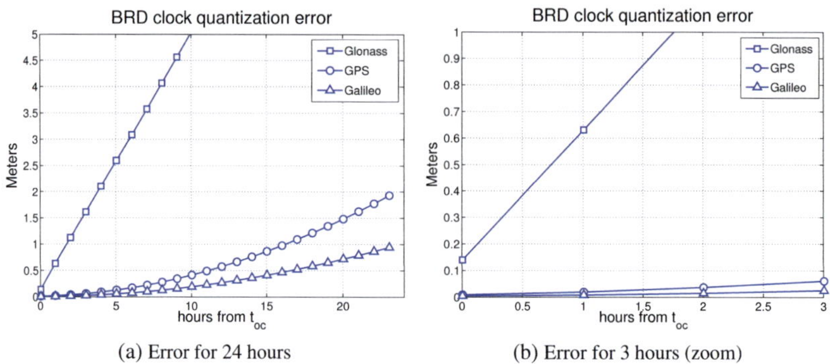

(a) Error for 24 hours   (b) Error for 3 hours (zoom)

Fig. 3.3: Contribution of the broadcast (BRD) clock quantization into the UERE

referred to the time of week. At least a one week range is allocated and the counter is short cycled at week transitions (restarted to zero). The new definition of CNAV data presents the only clock model optimization with respect to NAV data with a reduction of 5 bits and a different scale factor. GLONASS allocates only one full day; nomenclature is also different (t_b).

The clock model parameters (a_0, a_1 and a_2) are more similar, but with small differences in line with the type of clocks used by each system. Phase offset (a_0) has the larger allocation in Galileo (62 ms) in order to avoid time keeping operations in PHM mode. Bits allocation is lower in GLONASS with 1 ns quantization limit, only 11 bits for a_1 and no a_2 transmitted since only cesium clocks are employed

The transmission to the user of the clock model with less precision than the estimation due to the minimum representation imposed by the broadcast message generates an error into the range computed by the user. Figure 3.3 provides the translation of this error into the user equivalent range error (UERE). The so-called quantization error is computed by Equation 3.4 with the minimum values provided in Table 3.5 divided by two assuming a rounding function. If just the most significant bits are used in the quantization, instead of applying a rounding function, the minimum value can be directly used. As observed, the quantization is not a negligible error contributor for the system when long prediction times are targeted. Following the subfigure (a), a user applying a perfect predicted clock would have, after 24 hours ($t - t_{oc}$), a one meter error with Galileo, two meters with GPS and several meters with GLONASS.

The approach taken to reduce this error is the same as applied to the orbit prediction model. The navigation message is further divided into smaller intervals: 0.5 hours in GLONASS, 2 hours in GPS and 3 hours in Galileo, stored in memory and transmitted between contact times. As a consequence, the error for a user with the latest valid navigation message is the one between 0 and 3 hours as enlarged in Figure 3.3 (b).

3.5.4 Time keeping

When the satellite is injected in the final orbit after launch, the payload is normally switched off or in standby mode and needs to be powered on. Once active, the satellite local time counter starts to count from zero and the on-board frequency source will be offset by the initial frequency accuracy plus the relativistic effect. Time scales are defined by their origin and basic interval. The local counter has to be synchronized to the system time scale, in order to align the origin, and the atomic clock frequency synchronized to the system frequency as observed from the ground, in order to align the basic interval.

Satellite time needs to be initially synchronized to ground time and be kept afterwards within the navigation message limits documented in Table 3.5. The ground segment needs to perform both operations. Initialization can be also required after any operation which involves switching off the unit hosting the time counter or the reference frequency source. Since the frequency drift (a_2) integrates into (a_0) over time, the frequency drift maybe also be required to be steered.

In case the limits are reached, the intervention of the ground segment is required to remove the satellite of the constellation, to issue an event notice, to perform an intervention to adjust $a_0 - a_1$ values by telecommand, to reintroduce the satellite into the constellation and to finally set the message as valid. These interventions generate a heavy work load, which engenders possible sources of errors, and a drop of continuity and availability for the users (as further analyzed in Section 8.3.2). The impact on the system availability is directly linked to the number of adjustments needed per satellite lifetime. For this reason, a dedicated time keeping system and strategy was introduced in GPS Block-IIR in order to control the clock drift [46] at the price of a higher short-term noise. The overall timekeeping strategy for Block-IIA is well described in [46] and complemented for -IIR in [131].

Hereafter the strategy will be reviewed and extended. Three commands are normally envisaged to steer local ground clocks at timing laboratories. The same commands can also be used for satellites.

$d_{a0}(1s)$ Coarse phase adjustment. The time counter is adjusted to an integer second value. It allows steps of multiples of 1 second without modifying the 1PPS signal generation. Adjustment does not affect the signal generation but may affect the modulation of the navigation message as the message is synchronized with the system time. This command is only intended to be used at initialization with the navigation signals switched off.

$d_{a0}(1ns)$ Fine phase adjustment. The local 1PPS signal is delayed or advanced by a delta value which can be accurate to sub-nanoseconds. In theory, this adjustment can be performed with the signals switched on; however, it is normally applied off-line to avoid impact to the user and receiver tracking.

d_{a1} Frequency step adjustment. A frequency step is introduced into the frequency by a frequency synthesizer. Frequency steps are applied with the signal switched on but with the satellite declared out of service.

d_{a2} Frequency drift adjustment. This command is in practice performed by continuous d_{a1} steps to compensate for the clock drift.

Derived from these commands three timekeeping strategies exist depending on the commands used:

1. Use of only phase steps (d_{a0}). This approach is applied only in receivers as later presented in Figure 3.10.

2. Use of phase and frequency steps (d_{a0}, d_{a1}). This approach is used in GPS Block-IIA and GLONASS satellites. The satellite needs to be taken out of the active constellation. An example of the usage of these commands for time keeping is shown in Figure 3.5 where the timekeeping of SVN34 and SVN36 (Block-IIA) is performed with d_{a0} and d_{a1} adjustments. The phase is re-initialized and the frequency offset corrected once per year.

3. Use of phase of initial phase (d_{a0}) and frequency steps (d_{a1}) plus continuous frequency steering (d_{a2}). This strategy is used in Block II-R. After the initial phase and frequency adjustments, timekeeping is maintained with frequency drift d_{a2} adjustments avoiding the removal of the satellite from the constellation. An example is shown in Figure 3.5 for SVN57. This satellite was launched on 21st December 2007 and declared usable on 2nd January 2008. After the clock was activated, it was left free running until the zero frequency offset was reached around 1st May 2008. Afterwards, the clock was steered to cancel the drift observed in the previous period. The residual drift was monitored and further adjusted in April 2009. Frequency drift (a_2) steering is not typically used for Block-IIA satellites. Nevertheless, after the last adjustment in SVN34 and -36 beginning in 2010, both satellites were steered also in frequency drift which seems a new strategy of the enhanced Control Segment.

Timekeeping operations are visible in GNSS constellations by plotting the phase and frequency evolution over a long time span. Figure 3.6 shows the phase and frequency for the complete GPS constellation from 2008 till 2011 and reveals the GPS timekeeping strategy. The phase is left free running up to ± 0.8 milliseconds to keep long term predictions untill 210 days within ± 0.9 milliseconds envisaged in the message. The frequency of Block IIA satellites is kept within $\pm 2E\text{-}11$; while for block II-M and -F, it is kept within $\pm 5E\text{-}12$ - well below $\pm 1.9E\text{-}09$ allowed by the navigation limits. The frequency is adjusted mainly to steer the phase inside the limits.

The same analysis with phase and frequency offsets for GLONASS is shown in Figure 3.7. The timekeeping strategy is more difficult to identify due to the numerous operations required on GLONASS satellites and the lower drift rate associated to cesium standards. Phase and frequency seem to be initialized at arbitrary values. The phase offset is maintained by one order of magnitude within the $\pm$2.0E-3 seconds limits of the navigation message. The fractional frequency is kept two orders of magnitude below the $\pm$9.3E-10 limit specified for the message.

Despite the fact that no timekeeping strategy is publicly available for Galileo, from information on the navigation message definition and clock specifications it is possible to make some assumptions. Figure 3.4 presents a possible strategy based on an initial phase and fractional frequency offset synchronization to GST. Within the expected 12 year lifetime of the satellite, the time will require two maintenances with the RAFS and no action for the PHM. The number of re-synchronizations for the RAFS could be reduced by frequency drift steering da_2. In reality the clock drift can be lower than specified, as in GIOVE clocks, and this period may be extended.

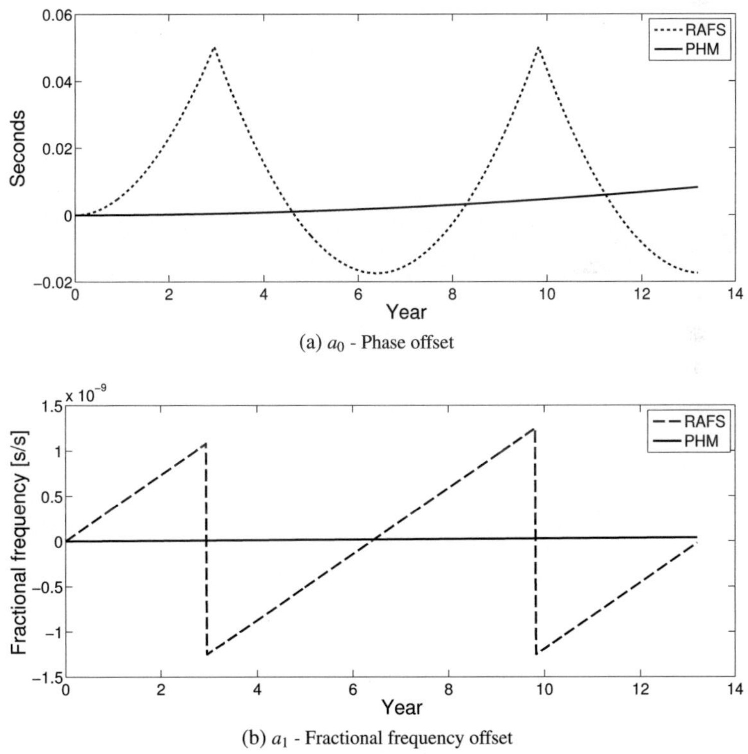

(a) a_0 - Phase offset

(b) a_1 - Fractional frequency offset

Fig. 3.4: Simulation of Time Keeping in Galileo

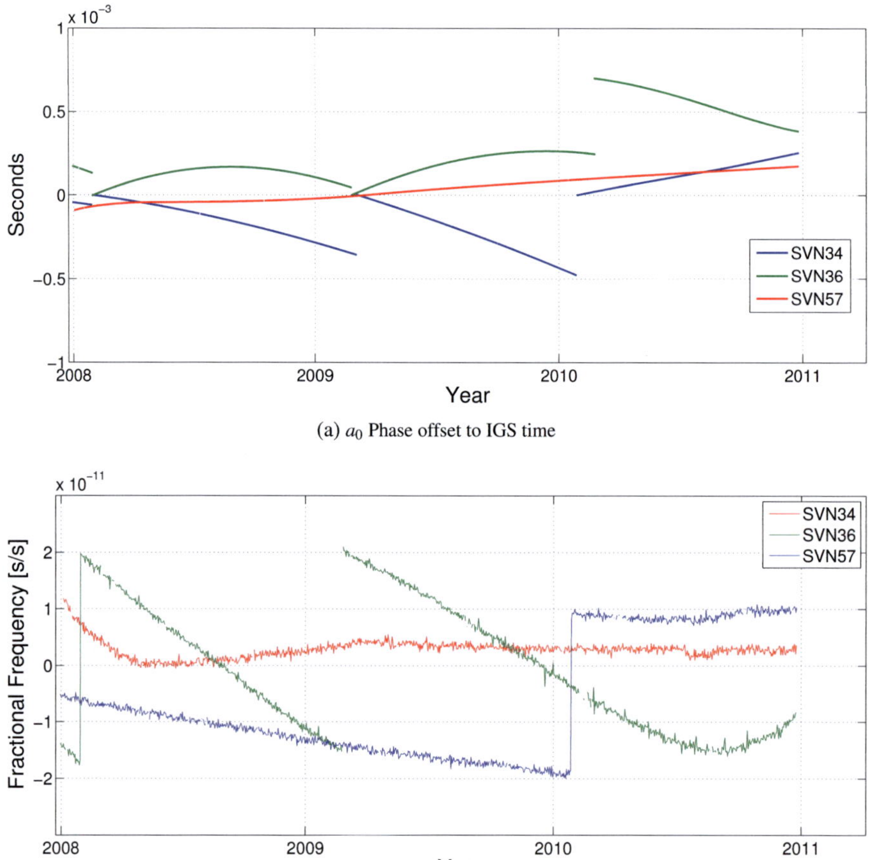

(a) a_0 Phase offset to IGS time

(b) a_1 Fractional frequency offset to IGS time

Fig. 3.5: Timekeeping strategy examples

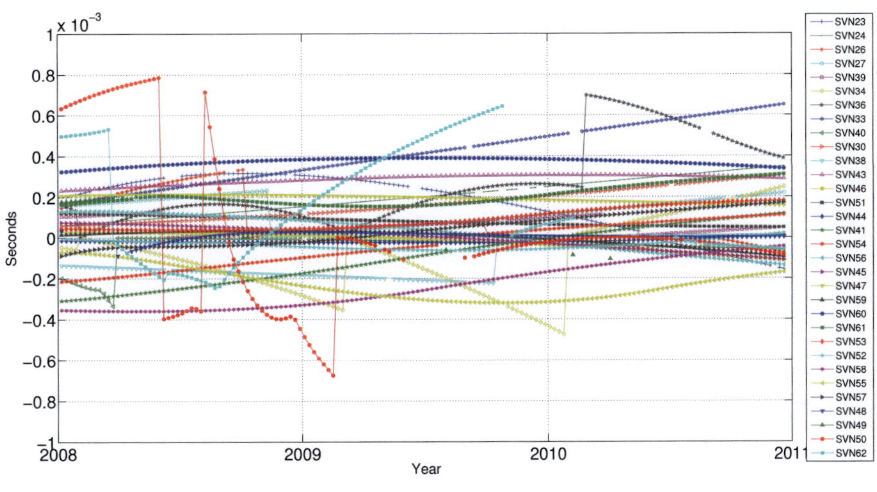

(a) a_0 - Phase offset to IGS time

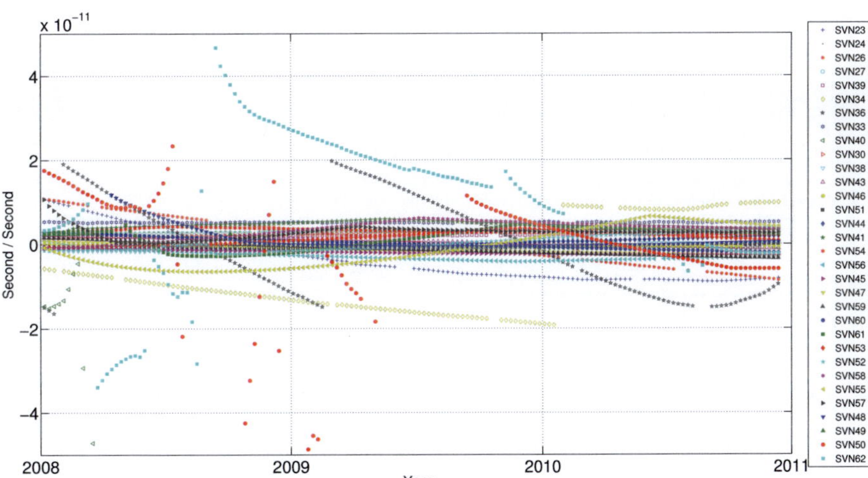

(b) a_1 - Fractional frequency offset to IGS time

Fig. 3.6: GPS timekeeping

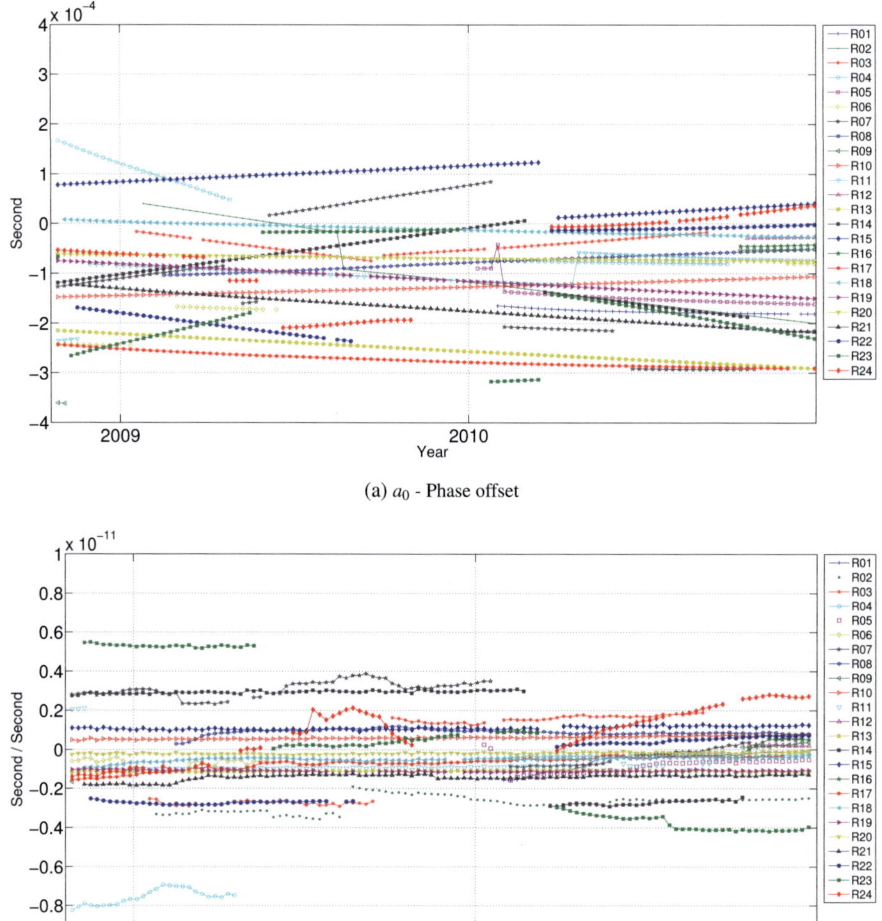

(a) a_0 - Phase offset

(b) a_1 - Fractional frequency offset

Fig. 3.7: GLONASS time keeping

3.6 Receiver time

3.6.1 Code and carrier phase measurements

In order to understand the receiver time, it is necessary to understand how the satellite time is recovered by the receiver. The pseudorange or code phase measurement can be considered as an absolute one-way time transfer between the satellite and receiver time. It is the basic observable element for navigation. By definition, the pseudorange is the difference between the time of reception (TOR) and the time of transmission (TOT) of the signal multiplied by the speed of light in vacuum (c), the receiver time being measured in the receiver time scale and the transmitted time measured in the satellite time scale [64].

$$PR(t) = (TOR(t) - TOT(t)) \cdot c \qquad [3.5]$$

The transmission time is generally recovered in GNSS receivers in sequential steps from three items of information:

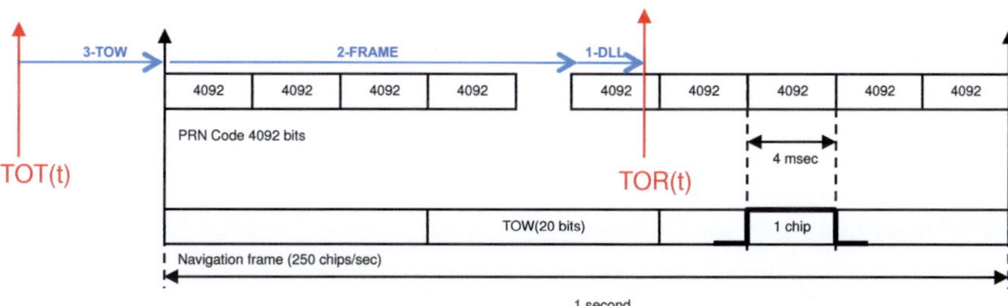

Fig. 3.8: Time of Transmission recovery in the Receiver for Galileo E1B signal

1. The tracking of the pseudo-random noise (PRN) code by delay lock loops (T_{DLL}) recovers the code phase with high accuracy, but with an ambiguity equal to the period of the PRN code length. This ambiguity is 1 ms for GPS C/A code and more diverse in Galileo (1,4,20 and 100 ms depending on the signal component tracked).

2. The detection of the navigation frame boundaries allows the extension of the ambiguity to the period of one navigation frame T_{FR} (e.g. for Galileo 1 sec in C/NAV, 2 sec in I/NAV and 10 sec in F/NAV).

3. The decoding of the time-tag fields in the navigation frames (T_{SYS}) enables a complete fix of the ambiguity, and hence also allow the receiver to obtain an absolute value of the transmission time.

The absolute time of transmission is composed by adding the three items of information:

$$TOT = T_{SYS} + T_{FR} + T_{DLL} \tag{3.6}$$

This three step approach is illustrated as an example in Figure 3.8 for Galileo E1BC signals. First, the correlation with the receiver replica by the DLL provides a high accuracy measurement with nanosecond accuracy (decimeter) but with 4 milliseconds ambiguity, since the same 4092 bits of code are repeated every 4 milliseconds. Second, if the navigation message can be decoded, the detection of the navigation frame boundary resolves the ambiguity to the 1 second order. Third, the recovery of the time-tag counters in INAV provides the final traceability to the local satellite time. In order to get the first PVT, the fourth step will be to decode the navigation message to get the ephemeris for orbit and clock corrections.

Consequently, the time to first fix (TTFF) depends on the time required by the receiver to get at least four valid pseudoranges to perform the PVT. The total time depends on the signal and navigation message design (as explained in section 3.5.2) and the receiver strategy to speed up some of the steps.

The code phase or pseudorange is an absolute time transfer from satellite to receiver time. The nanosecond accuracy provided by the code measurement can be extended to picoseconds order by measuring the carrier phase (also called accumulated Doppler). The carrier phase is an ambiguous time transfer measurement which needs the code to be resolved in an absolute way.

3.6.2 Time keeping in the receiver

Time of reception of the signal is obtained in Equation 3.6 as the instant where the correlation of the code replica in the receiver with the transmitted code is maximized. Measurement is performed in receiver time and it can be considered as an absolute time transfer from the satellite time to receiver time. Time at the receiver is driven by a local frequency source

The basic technology used for GNSS user clocks are crystal oscillators (XOs). For lower sensitivity to environment most of them are temperature-controlled (TCXO) and some are oven-controlled (OCXO). Major achievements in this domain have been the drastic reduction in power consumption and cost and major efforts have been devoted to miniature packaging with the ultimate goal of direct implementation on a single CMOS chip.

Whereas crystal oscillators provide a stable signal at short term, their time accuracy at medium and long term is affected by the aging of the crystal as well as environmental sensitivities. The use of atomic clocks clearly improves the medium-long term stability but with a higher price and power consumption. The atomic clock technology is typically limited to static sensors or some specific kinematic applications. However strong efforts have been dedicated over the last 10 years for the development of chip-scale atomic clocks with the size of a grain of rice, aiming to a faster acquisition by reducing the initial search space and a higher sensitivity by increasing

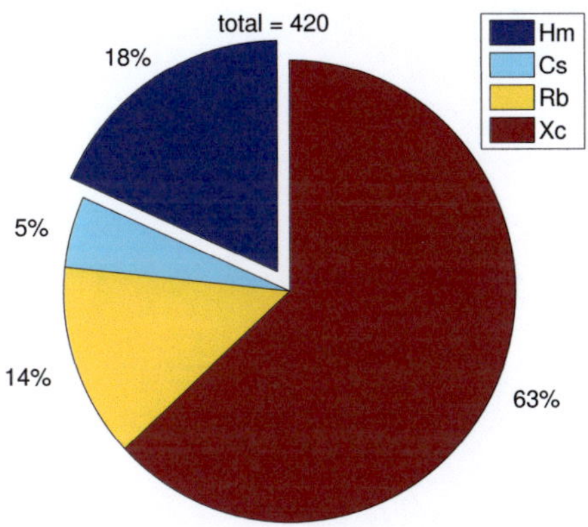

Fig. 3.9: IGS stations clocks

the coherent integration time in the receiver [81], the first commercial model being recently available [160]. Its use to increase GNSS robustness is also acknowledge by the US military Defense Advanced Research Projects Agency, which considerably promote this research area.

Sensor stations dedicated to time transfer in geodesy employ atomic frequency standards as e.g. in IGS network. Information on the log files of the sensor stations includes the type of clock used. Figure 3.9 shows the percentage of each clock type as extracted from the log files for the operating clocks on 01/01/2010. Around 37% of the receivers are connected to an atomic standard in the form of rubidium (Rb), H-maser (Hm) or cesium (Cs); the remaining 63% use a crystal oscillator (Xc).

Receiver time is reset after any station outage due to receiver, clock or any other problem at the station. First time synchronization is performed using the first PVT information or other external information. Subsequent timekeeping at the receiver is different for each type of technology. The exact solution depends on the type of clock technology used and the receiver manufacturer. In geodetic receivers the offset is left drifting within some limits till the phase is aligned by a phase step to the system time obtained from the PVT solution. Some crystal oscillators drift up to 1 second per day; another solution is applied in mass marked receivers by performing the measurement directly to the obtained GNSS time instead of receiver time.

In the IGS network during the analysed period from 2008 to 2011 only three of the stations with atomic standards present a continuous time scale without jumps or interruptions in the IGS final clock solutions. The three stations (USNO,USN3 and PTB) are located at time laboratories. The rest of the stations present a behaviour similar to GNOR station from the GIOVE network provided on Figure 3.10. The only difference is the different limits within which the phase is kept synchronized.

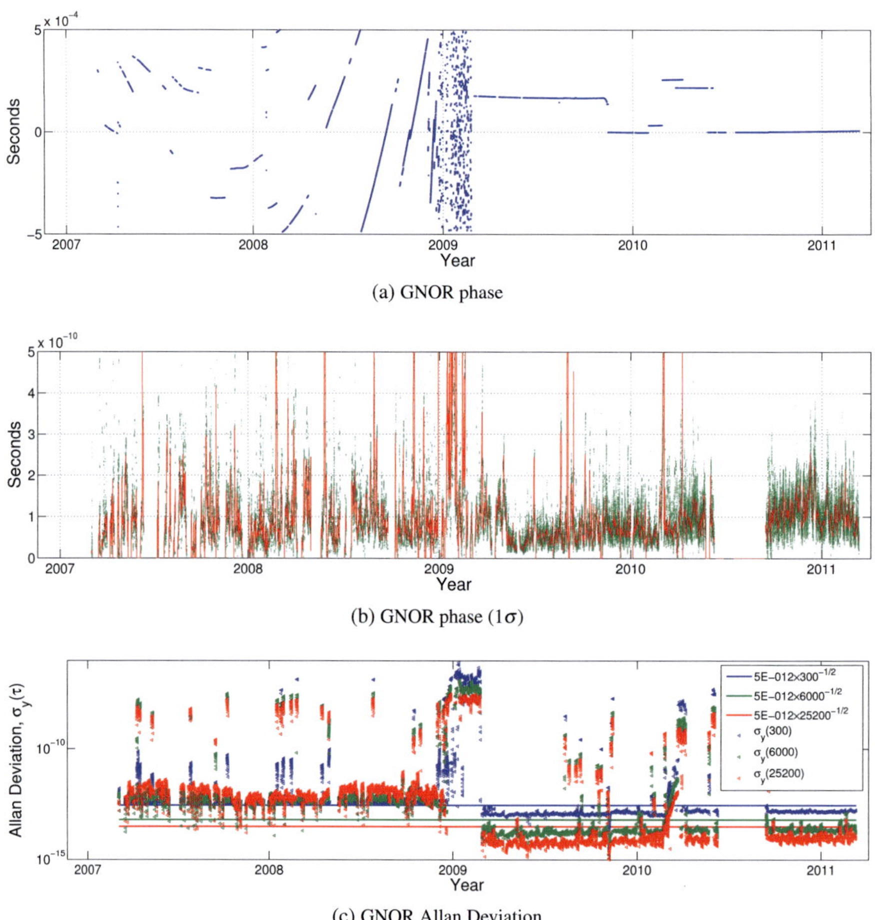

(a) GNOR phase

(b) GNOR phase (1σ)

(c) GNOR Allan Deviation

Fig. 3.10: GNOR station phase offset and standard deviation

GNOR is a good example as it operated three different clock types over the period from 2008 to 2011. The phase fluctuates within ± 0.5 milliseconds. Till beginning 2009 the clock operated on an external rubidium standard with a 1E-12 fractional frequency stability. The phase was adjusted several times per year to remain within the $\pm 0.5 ms$ limits. From January to March 2009 on the receiver operated on the internal crystal oscillator with a 1E-7 fractional frequency stability requiring a daily time steering of the clock. From March 2009 on, the station was connected to a H-maser with a frequency stability of 1E-13 at 1 second and 2E-15 flicker floor, the time evolution being only interrupted due to maintenances at the station or at the laboratory time distribution system.

The quality of the phase estimation for GNOR depends on the operated station clock, as observed from the estimated sigma. Each phase value in Figure 3.10(a) has been computed as the average value of 48 overlapping runs from which is possible to derive an associated standard deviation. Figure 3.10(b) shows the instantaneous standard deviation for each single value in red and a moving average in green in order to analyse its dependency of the operated clock. The observed averaged value is 0.1 nanoseconds for the periods operating with an atomic standard (rubidium or H-maser), while for the periods operating on the Crystal oscillator the standard deviation raises to 0.5 nanoseconds. This fact is due to the higher noise of the phase lock loop and carrier phase measurement due to the poorer stability. Although the clock phase is estimated every epoch, the estimation is affected by the higher frequency noise. A closer look to the figure shows that the standard deviation varies between 0.2 and 0.05 nanoseconds with several harmonic functions and trends of unknown source which seems to indicate some external sensitivity or POD residual effect. Receiver clock estimation seems to be an excellent indicator of the station quality as also demonstrated later in Section 6.3.2 when analysing the accuracy of geodetic time transfer.

3.7 Relativity in GNSS

All times involved in GNSS systems, satellite and receiver time scales, are affected by relativistic effects related to the invariance of the speed of light. The invariance of the speed of light implies that for each inertial frame the speed of light in vacuum c is the maximal speed of any signal or particle independent of the motion of the source. As a consequence the speed of light is also independent of the motion of the observer.

The numerical value of c has been defined by SI convention as the length of the path travelled by light in vacuum during a time interval of $1/c$ of a second [24]

$$c = 299792458 m/s.$$

The unit of space (meter) is linked to this definition. Several corrections for synchronizing

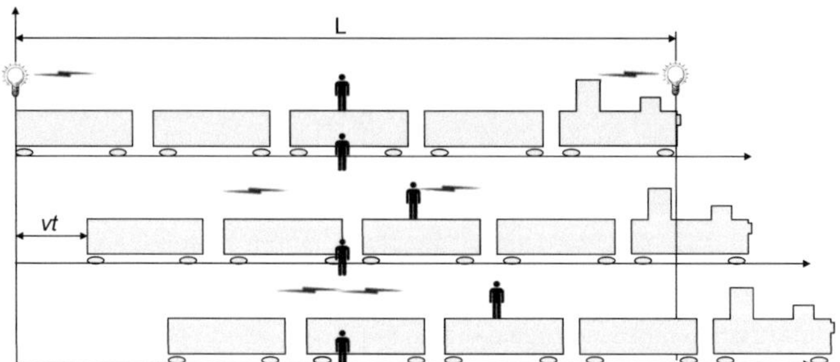

Fig. 3.11: Simultaneity

the GNSS clocks are derived from this principle of the invariance of the speed of light. The IERS conventions [104] and the on-going revision [73] provide an excellent summary of the relativistic effects currently applied to GNSS clocks for navigation and orbit determination. In order to better understand when and to which clocks they are applicable, it is appropriate to briefly introduce the basic principles behind each effect. Numerous publications by Ashby [7, 8, 9] describe in detail the principles applicable to GNSS. Even if the underlying principles are simple, a complete derivation can be complicated; in following subsections, according to Ashby, some simplifications are done to briefly introduce the corrections applied to GNSS time scales.

3.7.1 Sagnac effect

The Sagnac effect is derived from the Newtonian concept of simultaneity. Events which appear to occur simultaneously in one inertial frame may not appear simultaneously to observers in some other inertial frame, which is moving with respect to the first.

Consider, as illustrated in Figure 3.11, two events consisting of two light beams emitted from the two ends of a train of length L = 2x simultaneously as seen by two observers in the middle of the train, one static on the ground and the other on the train. The train is assumed to be moving to the right at speed v relative to the ground. The static observer will receive the two light signals in the middle point at the same time, while the moving observer will receive the light from the front first and will conclude that both light beams were not transmitted simultaneously, the light from the front being transmitted first. Since the speed of light must be invariant in each frame, the time relation between the two observers can be easily derived from the time of transmission of the signal in each frame.

For the moving observer, the time t' of light transmission on the front is

$$t' = -\frac{x}{c}$$

[3.7]

For the static observer the time for the light signal to arrive at the middle of the train is composed of the speed of light plus the train velocity

$$t = -\frac{x}{c+v} \approx -\frac{x}{c} + \frac{vx}{c^2}$$
[3.8]

The relation between the two time intervals is given by

$$t' = t - \frac{vx}{c^2}$$
[3.9]

This principle is quite useful in understanding the time synchronization related to a clock in the Earth-Centered, Earth-Fixed (ECEF) rotating frame with respect to a clock fixed in the Earth-Centered Inertial (ECI) frame. A clock on the Earth in the ECEF frame moves with a velocity $v = \omega r$ with respect to a clock in the ECI frame, where ω is the angular rotation speed of the Earth and r is the radius of the meridian containing both clocks. Applying Equation 3.9 the relationship between the clock in the resting ECI frame with time t, with respect to the clock in the moving ECEF frame with time t' will be given by:

$$t' = t - \frac{\omega r x}{c^2} = t - \frac{2\omega}{c^2}\frac{rx}{2}$$
[3.10]

The distance x between two clocks at the equator at a distance r from the center of the earth separated by an angle θ in radians will be $x = \theta r$. As a consequence the time difference when transferring time from Eastern to Western clocks all over the equator will be:

$$\Delta t = \frac{2\omega}{c^2}\frac{\theta}{2}r^2 = \frac{2\omega}{c^2}A_E$$
[3.11]

where $A_E = \frac{\theta}{2}r^2$ is the area of the sector of the circle enclosed by the time transfer process over the equator. In the case of a satellite and receiver clock synchronization process, the enclosed area A_E is determined from the two position vectors projected onto the earth's equatorial plane. The area of a triangle in a two-dimensional Euclidean space is given by $\frac{1}{2}|x_1y_2 - x_2y_1|$. As a consequence, when synchronizing satellite and receiver clocks, the following correction must be applied:

$$\Delta t = \frac{\omega}{c^2}\left(x_r y_s - x_r y_s\right)$$
[3.12]

3.7.2 Second order Doppler effect

The second order Doppler effect, also known as time dilatation, is also derived from the principle of the constancy of c. A clock in the moving frame beats more slowly than clocks in the resting frame to which it is successively compared.

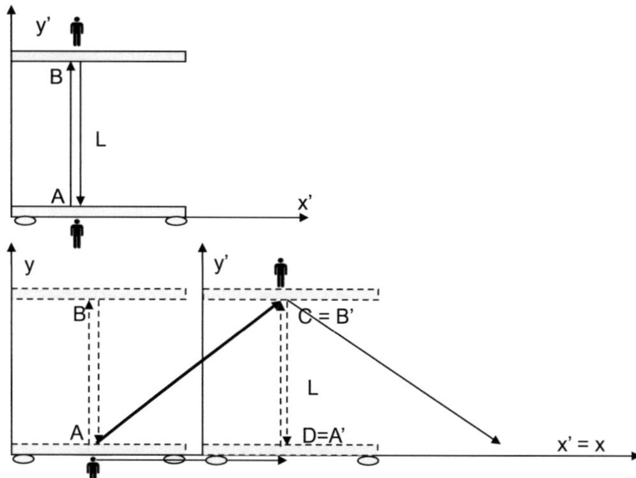

Fig. 3.12: Time dilatation principle

The relation can be established with a simple example. Consider the observers in Figure 3.12 -
one of them on a train moving to the right side, and which now carries inside the wagon a mirror
on the floor and another on the ceiling where a light ray is reflected vertically over a distance L.
At time zero, the axis of the static (x,y) and moving observer (x',y') are coincident and parallel.
The time t' required by the light to travel the distance L from A' to B' for the moving observer
is simply:

$$t' = L/c \qquad [3.13]$$

For the static observer, however, the point B has moved from B to C, and light follows the path
given from A to C. As a consequence of the principle of constancy of the speed of light, the
time elapsed is:

$$\overline{AC}^2 = \overline{AD}^2 + \overline{CD}^2$$
$$(ct)^2 = (vt)^2 + L^2$$
$$(ct)^2 - (vt)^2 = L^2 \qquad [3.14]$$
$$t^2(c^2 - v^2) = L^2$$
$$t = \frac{L}{\sqrt{c^2 - v^2}}$$

By substitution of the length $L = t'c$ from Equation 3.13 the following relation between the two
time intervals are finally obtained.

$$t = \frac{ct'}{\sqrt{c^2 - v^2}} \qquad [3.15]$$

As a consequence t' is

$$t' = \frac{t}{c}\sqrt{c^2 - v^2} = t\sqrt{1 - \frac{v^2}{c^2}} \qquad [3.16]$$

Since normally the ratio $\frac{v^2}{c^2}$ is small it can be approximated to the final expression

$$t' \approx t(1 - \tfrac{1}{2}\tfrac{v^2}{c^2}). \qquad [3.17]$$

3.7.3 Gravitational frequency shift

The gravitational frequency shift occurs when light signals are sent from one location to another with a different gravitational potential. Einstein's Equivalence Principle states that over a small region of space and time, a fictitious "gravitational" force induced by acceleration cannot be distinguished from a gravitational force produced by mass.

All experiments performed in a real gravitational field, such as in a laboratory on the surface of the earth where a gravitational field exists g, will have the same results as experiments performed in a laboratory in free space which is accelerated in the opposite direction with acceleration $a = -g$.

In consequence, gravitational fields can be reduced to zero by transforming them into a freely falling reference frame. The fictitious gravitational field caused by the acceleration then exactly cancels the real gravitational field. This basic principle is used by zero gravity experiments performed at drop towers where experiments are performed while free-falling inside a capsule, such as at the Center of Applied Space Technology and Microgravity (ZARM) facility in Bremen, Germany [169].

Let us imagine now an experiment where a plume and a 1-kg weight are located inside a capsule. Assuming that a perfect vacuum is created in the capsule and in the tower, the capsule is dropped and the experiment released. For an observer in the capsule reference frame, both objects appear to remain resting; while for the external observer, the objects fall with the capsule for the 110 meters of the tower length. Let us imagine the same capsule with a microwave transmitter, of a similar type as the navigation signal transmission used by GNSS, on the bottom and a receiver on the top. When the capsule is released, a wave is emitted to the receiver. The time required for the signal to propagate to the receiver in the capsule reference frame is

$$t = L/c \qquad [3.18]$$

while for an external observer during the propagation time the receiver has moved $dL = \tfrac{1}{2}gt^2$

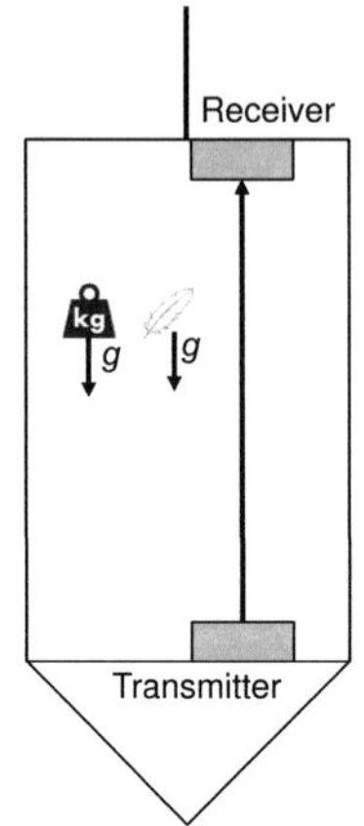

Fig. 3.13: Gravitational frequency shift

with a velocity

$$v = gt = gL/c \qquad [3.19]$$

and as a consequence the fractional frequency should be shifted by

$$\frac{\Delta}{f} = \frac{-v}{c} = \frac{-gL}{c^2} \qquad [3.20]$$

The quantity gL can be interpreted as the change in gravitational potential $\Delta\Phi$

$$\frac{\Delta}{f} = -\frac{\Delta\Phi}{c^2} \qquad [3.21]$$

Following [73] the potential for a clock A located at the geocentric reference system with the coordinate position $X_A(t)$ is

$$\Phi = +U_E(X_A) + V(X_A) - V(X_E) - x_A^i \partial_i V(X_E), \qquad [3.22]$$

where U_E denotes the Newtonian potential of the Earth at the position X_A of the clock in the geocentric frame, and V is the sum of the Newtonian potentials of the other bodies (mainly the Sun and the Moon) computed at a location X in barycentric coordinates, either at the position X_E of the Earth's center of mass, or at the clock location X_A.

The gravitational potential of the Earth U_E can be expressed as a series expansion in spherical harmonics. For frequency transfer, the contribution of third degree terms on the Earth's potential U_E and the tidal terms (the last three terms in Eq. 3.22) will be below 1E-15 in frequency and a few ps in time amplitude at the GPS orbit [177]. As a consequence, only the two main terms of

U_E need to be retained for current clock accuracies:

$$\Phi = U_E(X_A) \approx \frac{GM_E}{\rho} + \frac{GM_E a_E^2 J_2}{2\rho^3}(1 - 3cos^2\theta) \qquad [3.23]$$

where G is the Newtonian gravitational constant; M_E, a_E and J_2 are, respectively, the mass, the equatorial radius and the quadruple moment coefficient of the Earth and; ρ and θ are the radius and geocentric colatitude of the point of interest.

3.7.4 Periodic relativistic correction

Since radial distance and velocity are not constant, the effects on the satellite clock due to the gravitational frequency shift and second-order Doppler, vary according to orbit eccentricity. The correction can be derived from the integration of the higher order terms neglected in the previous section and can be described in a first order approximation by [9]:

$$\Delta t_r = \frac{2}{c^2}\sqrt{GM_E a}\, e \sin E \qquad [3.24]$$

Equation 3.24 can be expressed in a more convenient form without approximation by the following alternative Equation 3.25 where r and v are the position and velocity of the satellite at the instant of transmission.

$$\Delta t_r = \frac{2r \cdot v}{c^2}. \qquad [3.25]$$

Both equations use only the first term of the gravitational potential introducing a periodic error with an amplitude of 0.1 ns at half of the orbit period as suggested by [83]. This effect is also observed on GIOVE satellites - especially on the PHM where the lower noise allows the observation of a clear peak in the spectra at half the orbit period in Figure 7.9. The improved equation is also given by [83]:

$$\Delta t_r = \frac{2r \cdot v}{c^2} - \frac{a_E^2}{2a^2 c^2}J_2 \left[3\sqrt{GM_E a}\cdot \sin^2 i \cdot sin2u - 7\frac{GM_E}{a}\left(1 - \frac{3}{2}sin^2 i\right)t\right] \qquad [3.26]$$

where i is the orbit inclination, a is the semi-major axis and u the argument of latitude $u = (x+f)$, i.e., the sum of the true anomaly f and the argument of perigee x, and t is the GPS nominal time.

3.7.5 Measured values in orbit

In summary, the time and frequency comparison of moving clocks over long distances need to be treated in the context of special and general relativity. This is also the case for ground clocks spinning around the Earth's axis at a given height over the geoid and the satellite clocks orbiting the Earth.

symbol	value	unit	name	source
GM_E	3.9860044150E+14	m3/sec2	Geocentric gravitational constant	IERS 2003
c	299792458	m/sec	speed of light	BIPM
a_E	6.3781365500E+06	m	Earth's equatorial radius (tide free)	IERS 2003
J_2	1.0826267000E-03	-	Earth's dynamical form factor	IERS 2003
ω	7.2921151467E-05	rad/sec	Earth's mean angular velocity	IERS 2003

Tab. 3.6: Constants definition

The summary of major effects affecting satellite and ground clocks is provided in Equation 3.27 and the associated constants required to compute each value are provided in Table 3.6:

$$dt' = \left(\underbrace{\frac{GM_E}{c^2\rho}}_{1} + \underbrace{\frac{GM_E J_2 a_E^2}{2ac^2\rho^3}}_{2} + \underbrace{\frac{v^2}{2c^2}}_{3} \right) dt + \underbrace{\frac{2r \cdot v}{c^2}}_{4} + \underbrace{\frac{2\omega}{c^2}A_E}_{5} \qquad [3.27]$$

1. Earth gravitational contribution due to the Earth's mass, where ρ is a constant distance from the center of the Earth to the clock. The semi major axis a of the orbit is used for the satellite clock and the Earth's equatorial radius a_E for the ground clock. a_E is tide free, the total tidal effects have been removed with a model, as recommended by Resolution 16 of the 18th General Assembly of the IAG (1983) with quantities associated with the geopotential.

2. Earth quadrupole moment contribution.

3. 2nd order Doppler effect due to the clock velocity, where v is the Earth's spin velocity for ground clocks at the equator ($v = \omega a_E$) and the satellite velocity for a circular orbit ($v = \sqrt{\frac{GM_E}{a}}$) as first approximation.

4. Periodic relativistic correction. Since the orbit is elliptic an additional periodic correction needs to be included.

5. Sagnac correction.

Higher order terms are advised by IERS Conventions 2003 [104] for precise orbit determination. These effects need to be included when a high level of precision is required as for POD or Precise Point Positioning (PPP). Nevertheless, these terms are not retained here as they do not affect the conclusions and it simplifies the analysis.

The constant components (1)-(3) provide the net frequency shift to the fractional frequency to be observed after the satellite launch, the so-called factory frequency offset, which can be corrected from the ground segment by applying a frequency step adjustment at the initial time synchronisation (see Section 3.5.4); or by providing the offset in the a_1 term of the navigation

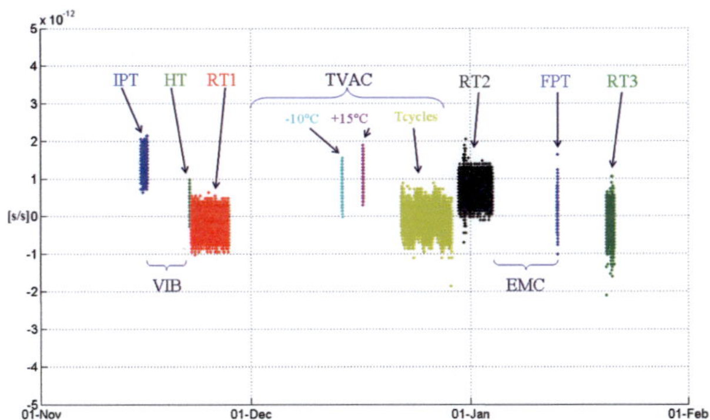

Fig. 3.14: Fractional frequency offset measured by the PHM during ground acceptance tests

message. The non-constant components (4) and (5) are periodic corrections with a magnitude of around 23 and 15 meters and a level of precision below 1 cm, left to be corrected by the user in the navigation algorithms.

Table 3.7 summarizes the constant components (1)-(3) in 3.27 for GIOVE-B (IERS2003), together with the values for GPS(WGS-84) reported in [124].

		GPS	GIOVE-B
	semi-major axis [m]	26561750	29551218
Satellite	Earth Gravitation	-1.670E-10	-1.501E-10
	Earth Quadrupole Moment	-5.211E-15	-3.784E-15
	2nd Order Doppler	-8.349E-11	-7.504E-11
Ground	Earth Gravitation	-6.953E-10	-6.953E-10
	Earth Quadrupole Moment	-3.764E-13	-3.764E-13
	2nd Order Doppler	-1.203E-12	-1.203E-12
SAT-GROUND	Net effect	4.465E-10	4.718E-10

Tab. 3.7: GPS and GALILEO relativistic effects

During acceptance tests on ground, the on-board clocks are subject to environmental tests (including vibration, thermal vacuum, etc.) during which it is possible to estimate the frequency repeatability of these clocks. It was confirmed that for RAFS, the repeatability was in the order of 5E-10. As a result, the RAFS could not be used to analyze relativistic frequency shift on-board the GIOVE spacecrafts. For PHM however, this is confirmed to be at the level of a few 1E-12.

Figure 3.14 depicts the fractional frequency of the PHM during the acceptance tests on ground: Initial Performance Tests(IPT), Vibration Tests (VIB), Health Test (HT), Reference Tests 1-2-3 (RT1-2-3), Thermal Vacuum Test (TVAC), Electromagnetic Compatibility Test

(EMC) and Final Performance Test (FPT).These measurements were obtained against an active hydrogen maser and the accuracy of these measurements is therefore expected to be at the 1E-13 level. This figure confirms that the frequency repeatability of the PHM is in the order of a few 1E-12 after satellite vibration.

In orbit, the fractional frequency offset may be estimated by POD techniques. Figure 8.4 presents the estimated fractional frequency offset of the PHM during its live time aboard GIOVE-B. The PHM frequency is estimated against a steered hydrogen maser located at USNO timing laboratory. The absolute frequency of the PHM is 10MHz which is changed later by a frequency synthesizer in the frequency control unit to the nominal value ($f_0 = 10.23$MHz) from which the navigation signals are derived. The frequency measured on ground was intentionally offset by -4E-10 from this value to account for relativistic effects to 10229999.99590920 Hz. Following the relativity theory the expected observed value in orbit should instead be shifted by +4.718E-10 to 10230000.00073570 Hz. Consequently, the expected frequency offset value in-orbit as observed from the ground should be the theoretic minus the precorrected value: +4.718E-10 - 4.00E-10 = 7.19E-11.

Over the first month of operation, when possible aging effects do not affect the validity of the results, the on-board PHM fractional frequency offset is estimated to be 7.75E-11, as opposed to an expected theoretical value of 7.19E-11 with respect to the measured ground frequency. Therefore, the PHM allows the measurement of the relativistic frequency shift with an error of 5.58E-12, corresponding to 1.2% accuracy with respect to the measured value. Table 3.8 summarizes the values.

	Frequency f_0 [Hz]	Freq. Offset Δf [Hz]	$\Delta f / f_0$ [s/s]	$\Delta f / f_0$ [s/day]
Nominal f_0	10230000.0000000	0.00000000	0	0
Ground	10229999.9959092	-0.00409083	-4.00E-10	-3.5E-05
Expected	10230000.0007357	0.00073574	7.19E-11	6.2E-06
In orbit	10230000.0007928	0.00079283	7.75E-11	6.7E-06
Delta		0.00005708	5.58E-12	4.8E-07

Tab. 3.8: GIOVE-B PHM frequency offset and relativity effect

3.8 Conclusions

This chapter has given an overview of how GNSS systems provide to the users an access to different time scales, from the international time scale creation to the final user receiver. Figure 3.15 provides a scheme of the traceability from Terrestrial Time (TT) creation to the provision to the user and the required transformations. Normal navigation users need only system time. Timing users require traceability to UTC which provides further access to other time scales used in Astronomy.

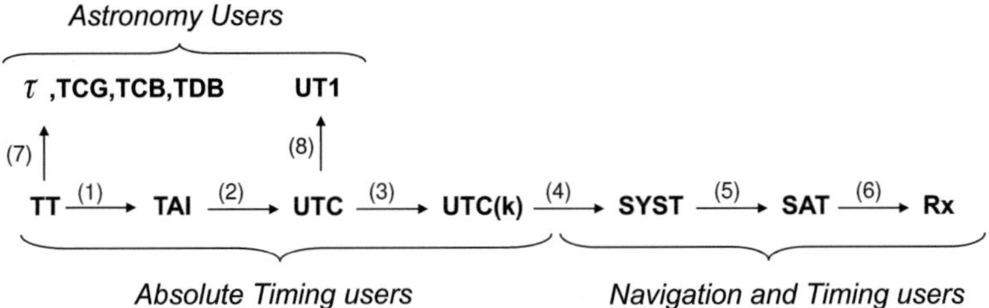

Fig. 3.15: Traceability between time scales in GNSS

Section 3.2 and 3.3 define the current atomic time scales based on the SI second definition as the basic interval. Elementary time scale is TT derived from the SI second definition on the rotating geoid and as a consequence represents the ideal time of a user on the Earth's surface. TAI is a physical realization of TT based on the measurement of atomic frequency standards distributed around the world. It is synchronized with TT apart from a constant offset (1):

$$TAI - TT = -32.184s \qquad [3.28]$$

TAI drifts slowly from UT1 based on the earth rotation as the earth rotation is slowing down. Even if most countries have some hours difference with respect to the solar time, UTC was introduced to follow UT1 ± 0.9 seconds requiring periodic integer ± 1 leap second steps corrections (2):

$$UTC - TAI = \pm LeapSeconds \qquad [3.29]$$

Section 3.4 explains how GNSS times are created as atomic time scales, maintained by the ground segment and linked to some UTC(k) creation. As UTC(k) contributes to UTC creation the traceability is provided by the BIPM on the CircularT (3):

$$UTC(k) - UTC = CircularT \qquad [3.30]$$

To support timing users the difference between UTC(k) and system time is provided by the navigation message of each satellite (4):

$$t_{sys} - UTC(k) = A_0 + A_1(t_{sys} - t_{oc}) \qquad [3.31]$$

It also has also to be remarked that other external service providers, as IGS or SBAS, may provide the traceability of the clock in step (5) to their own system time realization. As a consequence, the time solution achieved by the user will be referred to this time scale.

Section 3.5 explains how the satellite time is created on board, maintained within the message limits and transmitted to the user through the navigation message (5) :

$$t^{sat} - t_{sys} = a_0 + a_1(t_{sys} - t_{oc}) + a_2(t_{sys} - t_{oc})^2 \qquad [3.32]$$

The GNSS capabilities to provide traceability between the different time scales,in step (4) and (5), are particularly constrained by the navigation message specifications. The definition limits the minimum time to first fix, the maximum accuracy achievable in the time transfer and imposes a limit within which all traced time scales need to be synchronised by a dedicated timekeeping strategy applied from the ground.

Section 3.6 finally explains how the user recovers the satellite Time of Transmission (TOT) in the receiver through the decoding of the navigation message (time-tag information T_{sysT} and frames boundaries T_{FR} and the code delay at the Time of Reception (TOR) in the receiver (Rx). This measure is called pseudorange (PR) since the geometric range ρ also includes other contributions ξ (6) :

$$PR(t_{rec}) = TOR(t_{rec}) - TOT(t^{sat}) = T_{SYS} + T_{FR} + DLL = \rho/c + dt_{rec}(t_{sys}) + \xi \qquad [3.33]$$

In case the position is known, the pseudorange measurement can be corrected to remove the geometric range and the other contributions can then be modelelled or estimated to get the receiver time dt_{rec} offset to the system time (t_{sys}).

Finally, the definition of the time scales and the time transfer between moving clocks needs to be understood in the framework of the general relativity theory provided in Section 3.7. Time scales need to be understood in the reference frame and at the position in which they are defined. Astronomical users require that the time be referred to the Geocentric or Barycentric reference time scales (TCG and TCB). Transformation between TT and time scales defined in the geocentric or barycentric reference system (7) are provided in the IAU Resolutions and summarized in [104]. Satellite orbits are provided in the Earth's fixed frame; users of inertial frames may also be interested in the UT1-UTC difference and the Earth orientation parameters. This traceability (8) is provided by IERS or the new navigation messages of GLONASS-K and GPS-L5.

In GNSS, the time transfer between moving clocks requires 'relativistic corrections' in order to refer all clocks to the system time scale used as a reference. These relativistic corrections may be divided into two types: a first group creating a net frequency shift in the clock as observed from ground, and a second group producing a periodic variation to the mean value.

The frequency shift is normally compensated from the ground while the periodic contributions are corrected by the user. The initial clock frequency offset following its first activation in-orbit has an associated uncertainty, which is compensated from ground together with the relativistic shift. The superior frequency repeatability of the new clock technology provided by the PHM has allowed, within this dissertation, the measurement of the expected relativistic frequency shift (4.718E-10) with an error of 5.58E-12, corresponding to 1.2% of the measured value.

4 Timing signals realization

4.1 Introduction

The generation of any atomic time scale requires Atomic Frequency Standards (AFS). Special AFS are used aboard GNSS satellites due to the low mass, low power consumption and high reliability requirements. AFS represent the core element of the satellite time, being one of the technologies required for GNSS with limited flight experience in other satellites. This technology is currently only mature enough in some countries with a limited number of suppliers. The potential use by military systems restricts the exportability between countries. The availability of AFS technology by a diversity of reliable manufacturers is a key element for any autonomous GNSS system as demonstrated during GPS lifetime.

The atomic frequency standard signal is further modified by the other units part of the navigation payload before transmission to the user receivers. The name 'clock' is usually applied to the frequency standard on board the satellite even if it does not directly provide any time information. The term 'clock offset' is also used in the navigation message, or by IGS to refer to the difference between the ground and satellite time scales. However, the term 'timing signal' rather than 'clock offset' used by GPS performance reports [120] is more appropriate because the output of the atomic frequency standard is further modified by the electronics before being broadcast by the satellite and only the navigation signal includes time information.

Previous chapters introduced the different time scales in GNSS systems. This chapter intends to produce a more complete understanding of the satellite timing subsystem by examining the physical component, history, state of the art and future trends of its components. This understanding is absolutely necessary in order to explore the possibilites offered by the new AFS, signals and modulations offered by the upcoming GNSS satellites.

4.2 Atomic frequency standards

In 1967-68, Atomic Time was defined as being linked to the cesium transition [51] and formally adopted as the international system of units (SI) by the Bureau International des Poids et Mesures (BIPM) [24]:

The second is the duration of 9 192 631 770 periods of the radiation corresponding to the transition between the two hyperfine levels of the ground state of the cesium 133 atom.

Since 1983, the second has also defined the unit of length as the 17th General Conference on Weights and Measurements (CGPM) linked the length definition to the second [24] :

The metre is the length of the path travelled by light in vacuum during a time interval of 1/299 792 458 of a second.

Atomic clocks are instruments which, using a specific atomic transition, are able to deliver a signal in real time with the same frequency anywhere at any time, depending only on fundamental physical constants up to the limits of experimental error [10]. In Metrology the atomic clocks are named **Atomic Frequency Standards** (AFS) in relation to the frequency of the sinusoidal signal they supply. Following [75] the frequency standards can be considered *primary* in case that they provide a fundamental absolute reference measurement which does not need calibration or *secondary* standards if their value is assigned by relative measurements to a primary standard. In practice, all standards require traceability to TAI and the difference between primary and secondary depends on the required accuracy.

The actual definition of the SI second is linked to the cesium element. However, other atomic clocks exist which take advantage of cesium, hydrogen, rubidium or ionized mercury atoms. These atoms have an unpaired electron in the outer electron shell, the inner subshells being either full or empty. Under the effect of a suitable excitation energy, the atom can be shifted from its ground state into an excited state. The transition between these states occurs through emission or absorption of electromagnetic radiation. This electromagnetic radiation is used to tune a quartz oscillator which provides the reference frequency.

The physical package also depends on the associated electronic system used to generate the frequency. Two main technologies are used: *passive* or *active* frequency standards. Passive technology is the method used for all space clocks. The transition is excited by means of electromagnetic signals; and the closer the excitation frequency lies to the resonance frequency, the larger is the response. The probe frequency is generated by a quartz oscillator and frequency synthesis methods. The quartz is locked to the resonance signals using feedback loops. The H-maser can also be built as active. In its active form, the quartz is directly locked to the signal of the maser.

Space clocks were originally derived from ground clocks that were adapted to the space environment, as was the case for the first GPS Block-I rubidium. This origin has been maintained using only well proven technology used for space applications; nevertheless, the serious requirements for space also made this technology later transferable to ground applications. Currently, the main commercial AFS used for ground applications are rubidiums. Rubidium clocks rely on probing atomic vapour contained in small glass cells, and thus offer the advantages of the small overall size, low mass, and low power consumption, making them ideal candidates for many applications. They have the inconvenience of having a relatively high drift which makes necessary the synchronization to a primary standard. The second most commonly-used atomic clock on ground is the commercial cesium clock, based on magnetic deflection technology being a standard in all metrology and timing laboratories. The last ground technology used is the hydrogen maser. The active H-maser is the commercial frequency standard with the highest frequency stability for periods between 1 second and few hours.

Principal ground rubidium suppliers are Perkin Elmer (US) and Spectra Time (CH, formerly TEMEX). Symmetricom (US) is the main supplier of cesium clocks with other small suppliers using part of its technology. The main providers of Hydrogen Masers are Symmetricom (US), T4Science (CH), Vremya (RU) and Qvartz (RU). Even if other AFS are used at timing laboratories or are under development, only these types meet the reliability required for ground or space applications. All ground suppliers are direct or indirect GNSS suppliers by providing part of the physical package or associated electronics; however, space clocks are different in respect to the ground clocks for several reasons [171]:

1. Predictability of the time signal below the error budget associated by system design.

2. Reliable continuity of the signal provided over the up to 12 years life time assigned to the satellite.

3. Ability of the system operators to anticipate and prevent signal anomalies, based on limited telemetry of the clocks and ground monitoring of the stability through the L-band signal.

4. Programmatic issue of having adequate production source(s) of these devices with a limited commercial market.

Atomic frequency standards also are important ingredients to scientific missions in space - for example, the 'Gravity Probe A' mission flew a hydrogen maser to measure the gravitational redshift. The Huygens-Cassini mission to Saturn and its moon Titan employed two lamp-pumped rubidium clocks as frequency references for the Doppler-wind experiment. Still today, these two rubidium clocks are the only atomic frequency standards having left the Earth's orbit.

While GNSS satellite clock technologies benefit from improvement of their ground versions, the expansion of GNSS time hampers the evolution of new technologies on-ground and their later migration to orbit. As acknowledged in [171], in the early century the GNSS system has also been guilty of the lower development of the commercial AFS as ground and space applications usually prefer to introduce GNSS receivers in order to provide timing and positioning rather than include a dedicated AFS. Space qualified GNSS receivers are almost a standard

GNSS	Block	Type	Accuracy	$\sigma_y(1s)$	$\sigma_y(1day)$	$dF/F/^oC$	kg	V(l)	W	year
	I	Rb		6.0E-11	9.40E-14	2.00E-12	5.9	13.6	25	1
	I, IIA	Rb	5.00E-12		1.37E-13	1.00E-13	5.9	13.6	25	1
GPS	I, IIA	Cs	3.00E-12	1.0E-11	1.36E-13	5.00E-14	12.7	10.3	22	3
	IIR	Rb	5.00E-12	3.0E-12	1.50E-14	7.00E-14	5.3	4.5	15	10
	IIF	Rb	5.00E-12	2.5E-12	5.00E-15	5.00E-14	6.1	4.8	39	12
	IIF	Cs	2.00E-12	1.0E-11	6.00E-14	1.00E-13	15.1	12.7	33	10
	I	Rb								
GLO	I	Cs	1.00E-11	5.00E-11	5.00E-13	5.00E-13	39.6	83.3	80	1
	M	Cs	1.00E-11	2.00E-11	1.00E-13	2.00E-13	52.0	149.0	90	3
	K	Cs	1.00E-11	2.00E-11	1.00E-13	1.00E-13	32.0			
	K*	Cs	1.00E-11	1.00E-11	6.00E-14	5.00E-14	16.0			
GAL	I	Rb	5.00E-10	5.0E-12	3.00E-14	5.00E-14	3.4	2.6	35	12
	I	PHM	2.00E-13	1.0E-12	3.00E-15	3.00E-14	18.0	28	60	12

Tab. 4.1: GNSS clocks characteristics

equipment in all current low Earth observation missions and adaptations are also being studied for geostationary missions using the side lobes of the L-Band signals.

Current space AFS developments are reviewed within this section with a view to the basic technology, history of the space development, current status and future trends based on the same technology. The main characteristics are summarized in Table 4.1 which is based on a similar table in [124], reviewed and complemented with the bibliographies provided hereafter for each clock model. Selected metric values are the frequency repeatability accuracy after switch-on, frequency stability in terms of Allan deviation σ_y at $\tau =1$ second and 1 day, thermal sensitivity, mass, volume, power consumption and expected life time. Second source cesiums on GPS Block IIA and first source rubidium in GLONASS have been omitted in the table as limited information is currenlty available for these clocks. Finally, the most promising clock technologies envisaged as candidates for future GNSS space craft are briefly reviewed.

4.2.1 Cesium

Cesium clocks have a good long-term stability and low frequency drift which make them standard equipment in timing laboratories. The proven technology and low drift have aslo made them a clear choice for GNSS applications.

Figure 4.1 extracted from [18] shows the classical deflection cesium clock used in GNSS. Cesium atoms are emitted from an oven; then, three operations are performed: first, in a selection phase the polarizer deflects only the atoms that contain electrons in the ground state by applying a magnetic field; Second, in the excitation phase the electrons are shifted from the ground energy state to the next energy or hyperfine state. The stimulation takes place in a Ramsey ca-

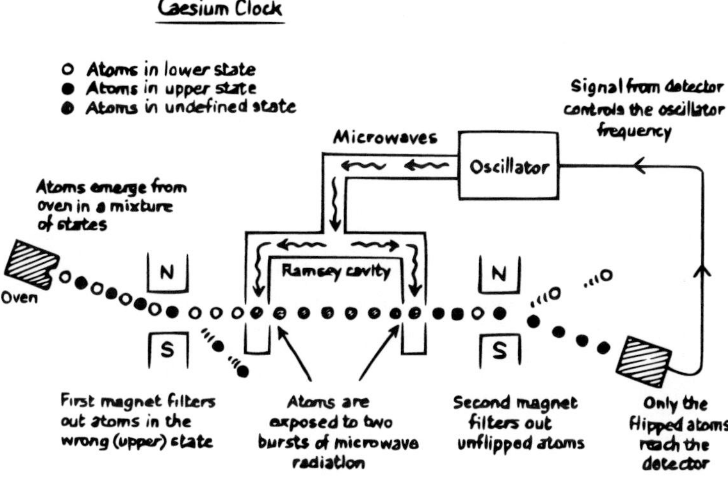

Fig. 4.1: Schematic representation of a cesium AFS using magnetic state selection. Source:NPL

vity, where the atoms are irradiated twice by a magnetic field close to the oscillating frequency (9.192.631.770 Hz). The closer the probe frequency, the higher the number of atoms which perform the transition to the hyperfine state. Third, in the detection phase, a second magnetic field is applied by a magnet(analyzer) which deflects only atoms which have made a transition to a hot-wire detector. Atoms are ionized and the ion current is proportional to the number of atoms deflected. The signal is used in a control loop to correct the Voltage-Controlled Crystal Oscillator (VCXO) providing the probe frequency used in the second step. The VCXO frequency is the actual clock output frequency of the device. A detailed description is available in [124, 18].

GPS

The Block I GPS satellites were foreseen to be equipped with one cesium clock (magnetic deflection technology) developed by Frequency and Time Systems (FTS) from existing ground technologies. However, this technology was not space-qualified at the start of the program. The first Block-I satellites carried only RAFS and Quartz technology. Only from NAVSTAR-4 onwards, one cesium clock was carried by each satellite [97]. The cesium clock was soon reported to meet the specifications [93], becoming the primary clock with respect to the rubidium.

Blocks II and IIA satellites were each equipped with two cesium clocks. Three potential suppliers were involved in the development of cesium clocks: FTS (now Symmetricom), Frequency Electronics Inc. (FEI) and Kernco. While FTS delivered 51 units for Blocks II/IIA, only 3 units were delivered by Kernco and 2 by FEI. The FEI clocks were launched on space vehicles 31 and 32. The Kernco clocks were launched on space vehicles 29, 30, and 34 [98].

On Block IIR satellites, it was originally intended to keep the same clock configuration as on Blocks II/IIA [108]. However, it was reported that, due to the interruption between the II/IIA and IIR programs, the clock suppliers were unable to maintain their know-how and qualification status. As a result, no cesium clocks were mounted on-board Block IIR/M although they were the preferred technology following the Block IIA experience [181].

Block IIF satellites are equipped with one cesium clock delivered by Datum-Timing (now Symmetricom) [44]. The first launched satellite from this block (SVN-62) carries the model 4415 S-Class following a newsletter from the manufacturer. The initial design was provided in [181], the final details being currently available on the manufacturer's website [159]. This clock was switched on briefly for testing in 2010 before changing to the rubidium and declaring the satellite operational; from what can be deduced, this clock is intended to be a secondary backup technology to the rubidium.

GLONASS

All clocks on-board GLONASS satellites are designed, manufactured and tested at the Russian Institute of Radionavigation and Time (RIRT). As for GPS, there have been various generations of clocks [60]. In the first operational phase (GLONASS Blocks IIa,b,c), the spacecraft were equipped with 3 cesium clocks of magnetic deflection technology ('GEM'), included in an integrated 'space-borne time/frequency standard' (STFS) system. Only one clock is operated at a time, the other two being cold redundant [15]. The modernized version of GLONASS (GLONASS-M) also includes 3 cesium clocks of magnetic deflection technology ('MALAKHIT') with improved performances and extended lifetime. Little information exists on this clock family. Some improvements with respect to previous publications [60] have been reported in RIRT presentations at ION conferences concerning the mass (32 kg instead of 54 kg) and a factor 2 in the short term stability and thermal sensitivity [16]. The same presentation reported efforts devoted to further improving the performance of the current cesium clocks for GLONASS-K - in particular related to major mass reduction (down to 16 kg), lifetime and manufacturing reproducibility. Confirmation of the final development of the clock is not yet available; nevertheless, the latest RIRT presentations state that such a performance has been achieved on the first GLONASS-K satellite [142].

New developments

It is possible to replace the twofold magnetic selection by interaction with laser fields. Optical preparation and detection have been employed in ground primary standards since the late nineties. Early on in the on-board clock development process for Galileo, it was felt that a third piece of technology could represent risk mitigation in case of major development issues, in particular with the PHM that had previously had no flight experience. Based on parallel investigations performed in the late eighties both in Switzerland (Oscilloquartz) and in France (Observatoire de Paris), it was recognized that the optically-pumped thermal cesium beam technology could provide an interesting alternative solution, with stability, mass and power consumption lying in-between the ones of RAFS and PHM. The selected cesium technology was different from the one used in GPS and GLONASS on-board clocks or in current commercial cesium clocks that rely on magnetically deflected thermal cesium beam technology, which has intrinsically poorer performances and a higher mass. In the early years of the new century, the interest in optically-pumped technology was confirmed by a number of studies and pre-developments both in France (at Tekelec Systèmes and Alcatel Space, now Thales Alenia Space, F) and in Switzerland (at Observatoire de Neuchatel). In 2008, a feasibility study of an on-board cesium clock for Galileo was conducted, led by Thales Electron Devices (F) and combining both the French and Swiss teams demonstrated with a prototype that the expected performances could be reached with a very elegant and simple solution taking the best of the previous developments and studies [71].

Symmetricom was also involved in the development of optically-pumped cesium beam technology based on the model on-board Block IIF [95]. Development was reported to be continuing in 2007, in particular through the use of European laser diode technologies from Eagleyard. Since then, the actual status of this development has remained unclear as this technology is no longer reported as being considered for GPS III.

The flight time in the microwave cavity is a key factor in cesium clocks. This time can be increased by the use of cesium fountains up to the limits imposed by the Earth gravity. The use of a cesium atomic fountain in space was soon identified as an advantage to increase the flight time in the microwave cavity. The PHARAO clock (from french Projet d'Horloge Atomique a Refroidissement d'Atomes en Orbite) has been designed with this objective and it is expected to be launched as part of the ACES ensemble in 2013. It will have an unprecedented flight stability (with $\sigma_y = 7 \times 10^{-14} \tau^{-1/2}$) really interesting for GNSS application. However, with a power consumption of 114 W and 91 kg mass, the technology lies well above the mass and power budget used in GNSS.

4.2.2 Rubidium

Figure 4.2, extracted from the National Institute of Standards and Technology (NIST) website on 'http://tf.nist.gov/general/enc-re.htm', shows the classical rubidium clock used in GNSS. In the rubidium AFS, the same three operations are performed as in the cesium AFS. First, in a selection phase, a light beam is generated by the rubidium lamp. The light passes through a cell filter with the rubidium isotope 85. The filter allows only two visible frequencies to continue. Second, when the beam reaches the resonance cell, the outer electrons of the isotope ^{87}Rb are

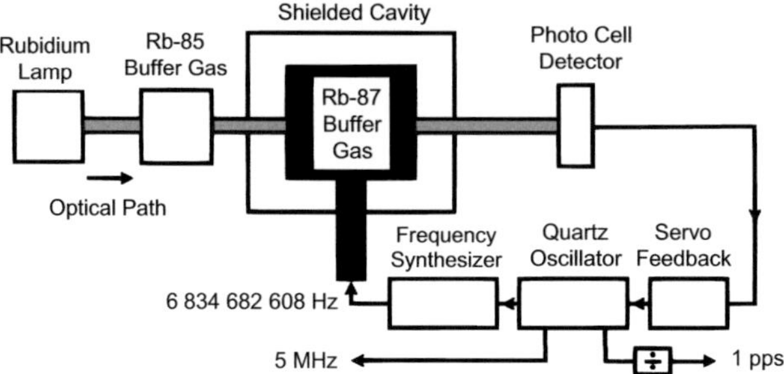

Fig. 4.2: Schematic representation of a rubidium cell AFS. Source: NIST website

excited to the hyperfine transition which receives a microwave interrogation signal. This increases the absorption of the light beam. Third, in the detector, the absorption decreases the output current at the photo detector. The minimum occurs at the maximum of electron transitions. The signal is used in a control loop to correct the VCXO which then provides the probe frequency used in the microwave cavity. The small size required for the resonance cell allows it to achieve a small physical package with low power consumption and price. This technology is the favourite for mass market applications on ground. In addition, this is a wellproven piece of technology making it a favourite candidate for GNSS payloads.

GPS

GPS satellites were the first spacecraft to fly a rubidium clock. The Block I GPS satellites were equipped with 3 rubidium clocks adapted from existing ground technologies. Although different atomic frequency standards were available on the commercial market, only rubidium standards could meet Air Force space qualification and be set into production quickly enough to meet the planned launch date of the first Block I satellites in February 1978. The GPS space-based rubidium atomic clock technology was derived from a unit produced by Efratom, a small company initially based in Germany [82]. The company lacked space experience and the space qualification was achieved with Rockwell's Autonetics, now Boeing Anaheim [123]. The first Block-I satellites embarked only rubidium clocks and, from SVN-8 in 1983, one cesium clock was embarked by satellite [40].

Blocks II and IIA were each equipped with 2 rubidium clocks. In total 56, units were delivered from the same supplier as for the Block I. Some improvements were introduced in the clock from flight experience, especially on the thermal control. Overall, the rubidium family associated to Block-I and -II/IIA had a better performance over one day than the cesium variety, and therefore were preferred for navigation; but the higher operational temperature, thermal sensitivity, drift and the lower reliability made the cesium the first choice for operations. In particular, the initial short 1 year life time of the rubidium was identified as being a major drawback with respect to the cesium [40, 108].

On Block IIR, it was originally foreseen to keep the same clock configuration as on Blocks II/IIA with the cesium as the primary standard [108]. However, it was reported that, due to the interruption between II/IIA and IIR programs, the cesium clock suppliers were unable to maintain their know-how and qualification status. As a result,

Block IIR/RM, due to problems in the qualification of the cesium clocks, embarked only RAFS coming from a new supplier (EG&G,now PerkinElmer). EG&G proposed a design similar to Block-I which was not finally retained to go into production [144]. Several prototypes were created from the first proposal until finally arriving at the final embarked design [145]. This clock is performing extremely well in orbit [47], as well as on ground. If reliability was an issue for the first Block-I,-II RAFS, the drawback was resolved with the new clocks. The GPS operations squadron reported only one failure on these IIR RAFS (for SVN61) since the launch of the first IIR spacecraft in 1997 untill 2008 [143]. For the three year period between 2008 and 2011 analyzed in this dissertation, only one event was observed associated also to SVN-61 (see Figure 8.8).

Block IIF is equipped with two RAFS from PerkinElmer based on the Block IIR design enhanced by the introduction of a Xenon lamp buffer gas and a thin-film spectral filter in the physics package, with significant improvement to the medium to long-term stability and signal to noise ratio [44, 127].

The same manufacturer will design and implement several engineering advances into its heritage Block-IIR and -IIF model, as well as qualify and deliver flight units for the first two Block IIIA satellites following the award of the contract in 2009 [62]. Together with the replacement of obsolete electronics in its heritage model, some engineering advances have been presented for a new flight unit which will be qualified and embarked on the first two satellites. The output signal will be provided at the 'natural frequency' 13.4MHz of the clock instead of the 10.23MHz in Block IIF due to the removal of the synthesizer, with a gain in volume, weight and short term noise (1E-12$\times \tau^{-1/2}$, [44]).

GLONASS

In the pre-operational phase of GLONASS (Block I, 1982-1985), all spacecraft were equipped with 2 rubidium clocks ('BERYL') with a design lifetime of one year, and stability reported to be 5E-12 at one day interval [98]. From 1985, on only cesium clocks are used. The lower mass GLONASS-K (Uragan-K) satellites were expected to have both rubidium and cesium AFSs. The first GLONASS-K launched in 2010 with new Code Division Multiple Access (CDMA) signals seems to carry only cesium standards considering the drift and noise being in line with previous models.

Galileo

Rubidium is the second baseline on-board clock technology for Galileo. Based on lamp-pumped vapor-cell technology, its development started in the early nineties at the Observatoire de Neuchatel (CH) originally for a Russian space Very Long Baseline Interferometry (VLBI) mission (RadioAstron). The development was taken over by Tekelec Neuchatel Time (TNT, CH, later Temex time and now Spectratime) for Galileo and the development steps included an industrialization contract resulting in a first Engineering Model. The qualification was performed together with Astrium GmbH (D) who was in charge of the electronics. Six flight models were delivered in the frame of the GIOVE program (2 for each GIOVE satellite and 2 spares) [146] and 8 flight models have been delivered for the in-orbit validation phase. Also two RAFS per satellite are envisaged for the fourteen initial FOC-1 satellites manufactured by OHB System AG [97, 98].

COMPASS,QZSS,IRNSS

The early Beidou satellites did not contain any atomic frequency standards and China lacks atomic clock technology that can survive the harsh space environment. To compensate, China purchased rubidium atomic clocks from the Swiss company Temex (now Spectratime). Compass M-1 was launched before the delivery of the European clocks [98, 43]. Three Chinese RAFS were supposed to be embarked on the first medium Earth orbit satellite (M-1) launched in 2007. The performance of the time signal on-board M-1 is reported to be worse than the GPS or Galileo timing signal [68].

Spectratime under contract with Astrium GmbH (D), will also deliver RAFS to the Indian Regional Navigation Satellite System (IRNSS) [98].

The quasi-zenith satellite system (QZSS) is a GPS augmentation system for Japan. The first satellite (QZS-1 or MICHIBIKI) was launched on September 11th, 2010. Despite efforts to embark a dedicated Japanese clock, the first satellite carries GPS standards as acknowledged on the Perkin Elmer website.

New developments

As early as in the nineties, it was identified that the use of laser diodes instead of discharge lamps in the RAFS would have several advantages in terms of both performance and operations [110]. Since then, a large effort has been dedicated in laboratories worldwide, with the aim of reaching PHM stability performance with a package as compact as RAFS. Various clock schemes, configurations and experiments have been performed with few notable successes.

In Europe, a development activity started in 2001 with Observatoire de Neuchatel (CH) to investigate the continuous double-resonance clock scheme. This activity demonstrated that with this scheme the short-term stability of the PHM could be reached and also lead to the development of extremely compact stabilized laser heads. In 2005, a follow-up activity was started to develop the key building block technologies and demonstrate both the short, medium and long-term stabilities. Early results have recently been reported [2], including the implementation of passive laser noise cancellation techniques.

Parallel investigations started in 2003 with the Italian 'Istituto Nazionale de Ricerca in Metrologia' on various clock schemes, including coherent population trapping with maser detection [92], pulsed optical pumping with maser detection and pulsed optical pumping with optical detection [109].

Russian RIRT is also involved in the development of low mass rubidium clock technologies based both on lamp pumping and laser pumping [16]. No detailed test results are available however.

4.2.3 PHM

Figure 4.3, extracted from the NIST website 'http://tf.nist.gov/general/enc-h.htm', shows the original construction of the hydrogen MASER (Microwave Amplification by Stimulated Emission of Radiation). In the first step, hydrogen gas is emitted in a beam and the states are separated by magnetic fields. In a second step, the atoms enter a storage bulb surrounded by a resonant cavity tuned to the frequency of the atomic transition. The microwave signal generated is used to lock a quartz oscillator. In the smaller passive version, the resonance cavity is supplied by a probe frequency at the resonance frequency, the maser amplifies the signal; and then a third step is introduced to lock the VCXO, thus providing the input frequency to the maximum range of the signal.

Galileo

The Passive Hydrogen Maser (PHM) is one of two baseline on-board clock technologies for Galileo. Its development started in the late nineties with the Observatoire de Neuchatel (CH), Galileo Avionica (I, now Selex Galileo) and Tekelec Neuchatel Time (CH, now Spectratime). After an industrialization phase during which the design was validated, a qualification phase resulted in the delivery of four qualification models [146]. Selex Galileo delivered two flight models to be embarked on-board GIOVE-B and eight flight models for the four IOV spacecraft [19]. Also two units per satellite are envisaged for the FOC-1 satellites manufactured by OHB.

Due to its excellent and unrivalled stability (less than 1 nsec of accumulated time error over 1 day), the PHM is considered to be the primary clock on-board the Galileo satellites. The PHM is the most stable AFS in orbit in terms of performance specifications as covered in Table 4.1.

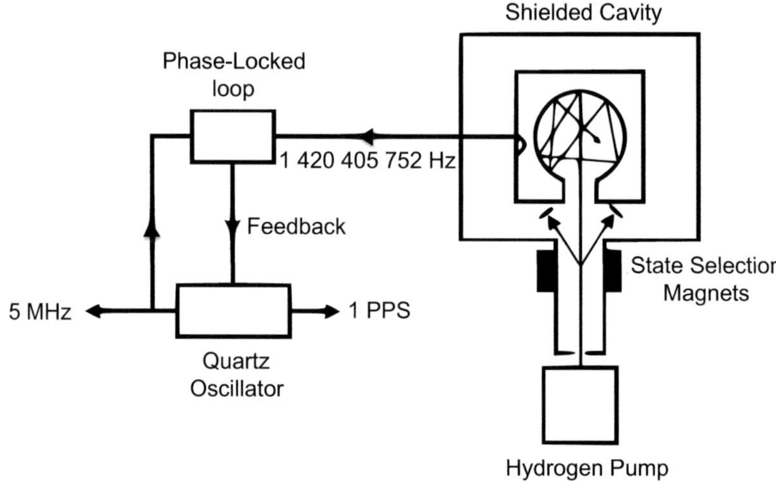

Fig. 4.3: Schematic representation of an active hydrogen maser. Source: NIST website

New developments

In the eighties, efforts at the US Naval Research Laboratory (NRL) were dedicated to the development of an on-board hydrogen maser for space applications [172]. This development resulted in a breadboard with performances similar to the European PHM [102]. Further industrialization and qualification steps have not been materialized and no further development of this technology for GPS-III has been reported.

Russia is one of the main suppliers of active H-maser technology. Some developments of both on-board passive and active hydrogen masers in Russia have also been reported. A breadboard model and test results have been presented by RIRT [61]. Similarly, the Russian company Vremya, commercializing ground passive and active hydrogen masers, reported by press release in 2005, to be studying the adaptation of their ground design to space level for both navigation missions, as well as for a Russian space VLBI mission (RadioAstron).

The hydrogen maser has also been reported to be developed by the Japanese National Institute of Information and Communications Technology (NICT) in collaboration with Anritsu Corporation for QZSS [77]. However, the first satellites carry rubidium AFS as primary clocks and no further development have been reported from 2007 on [76]. Despite being the best GNSS performing clock in orbit, the PHM is the heaviest equipment of the Galileo payload. It was identified that the design of the microwave cavity of the PHM (that makes up most of its volume) could be notably reduced with only minor impact on the stability performance. An activity was initiated to design, manufacture and test a physics package with reduced size, and then integrate it with a modified electronics package. In parallel, the design of the electronics package is being reviewed and updated in order to accommodate this reduced physics package. Preliminary outcomes on the physics package show results in line with the expectations with a mass reduction from 18 to 12 kg [20].

4.2.4 Hg+

Hydrogen, rubidium and cesium atoms are members of the first group of the periodic table, the so-called Alkali metals. These metals have only one electron in their outer shell which is used to create the hyperfine splitting of the ground state. Ions can be also confined in vacuum and used to build instruments that can be considered as reflecting atomic time and frequency standards.

The Jet Propulsion Laboratory already proposed in early nineties the Hg+ as an alternative to cesium standard [137]. Currently, the Jet Propulsion Laboratory (JPL) is investigating a mercury atomic frequency standard, which is a mercury ion storage clock for future GPS use. Spectra Time with other institutional partners and financial support from the Swiss space office is also expected to start investigating a mercury ion storage clock for future Galileo missions [98].

4.2.5 Optical clocks

The last topic of intensive research and development activities is in the field of optical clocks. It was long identified that the use of atomic transitions in the optical domain (as opposed to the microwave domain as used in all clocks described so far) would bring several orders of magnitude improvement in terms of frequency stability and accuracy. Over the last few years, all national metrology laboratories and a number of research institutes have embarked on the search for the best clock configuration and atomic species, with results surpassing the stability and accuracy of the best atomic clocks based on microwave transitions. This was made possible thanks to the advances in particular in the field of high resolution optical spectroscopy, laser cooling and trapping of atoms and ions, ultra-stable lasers and optical frequency combs.

On-ground, it has been demonstrated in various laboratories that optical clocks can indeed achieve stability and accuracy performances never reached before (down to the 1E-18 level), with various types of atoms or ions, in various configurations and using various clock schemes. This has opened the room to a multitude of research and developments in the field of optical spectroscopy, optical metrology or fundamental physics. From a time metrology point of view, several transitions in the optical domain are already considered as a secondary representation of the second. In the near future, once the optical clocks reach the level of operational reliability of the current primary standards, and once the time and frequency transfer techniques have improved their performance, it is likely that the second will be defined based on a transition in the optical domain. This technology is not expected to be available in the medium term for

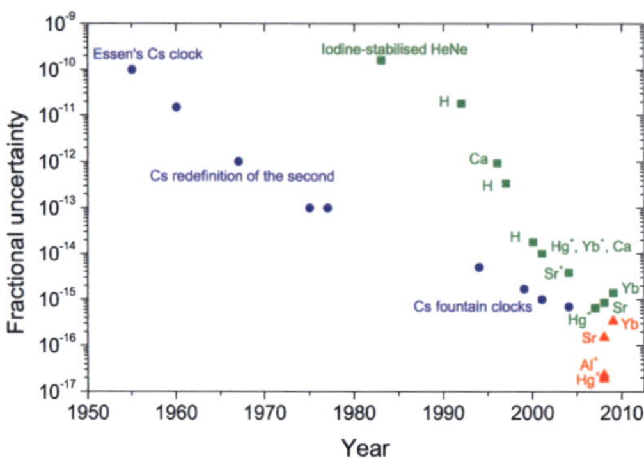

Fig. 4.4: Cesium microwave AFS (blue circles) versus optical frequency standards (green and red squares). [99]

GNSS systems. Major interest is in the research field and it is strongly proposed to be flown in a fundamental physics mission [49], opening the way for a future use in GNSS payloads.

4.3 Frequency distribution unit

All GNSS use a frequency distribution unit (FDU) which, together with the AFS, composes the timing subsystem. This unit has three basic functions. First, it operates as a switch between the different clocks to select one as the nominal clock in charge of providing the frequency for the navigation chain. Second, it changes the nominal clock frequency output of the different clock technologies on-board (e.g. 10.0028MHz or 10 MHz in Galileo) to the reference frequency (10.23Mhz) signal for the navigation generation unit and other platform equipment. Third, the unit is also in charge of adjusting the frequency output using a direct digital synthesizer commanded by ground operations to keep the satellite time inside the navigation limits (as explained in Section 3.5.4). Besides these basic three objectives, other functionalities can be implemented increasing the complexity of the unit, depending on the satellite design and objectives.

Figure 4.5 presents an overview of the design in GPS satellites. Early GPS Block-I satellites already included a frequency synthesizer as can be observed in the payload diagram in [78]; verification of the frequency adjustment functionality for this Block-I unit is also observable during the steering of the SVN-1 time signal while operating with a Quartz clock [105, Figure 5]. In Block-IIA, the unit was extended with the addition of Selective Availability capabilities. Block-IIR introduced a more complex design, the frequency synthesizer and distribution unit (FSDU) being used in Block-IIA was replaced by the Time Keeping System (TKS). The simpler design from Block-IIA was later recovered for Block-IIF [55]. Finally, the FDU to be carried in GPS-III will permanently remove SA capability as announced by the DoD (Release note number 1126-07).

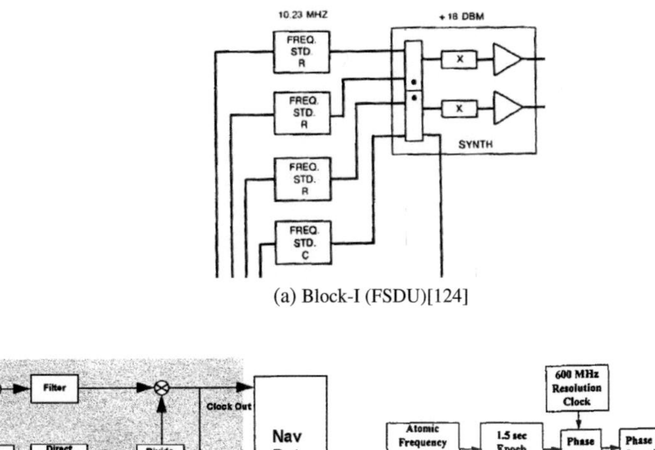

(a) Block-I (FSDU)[124]

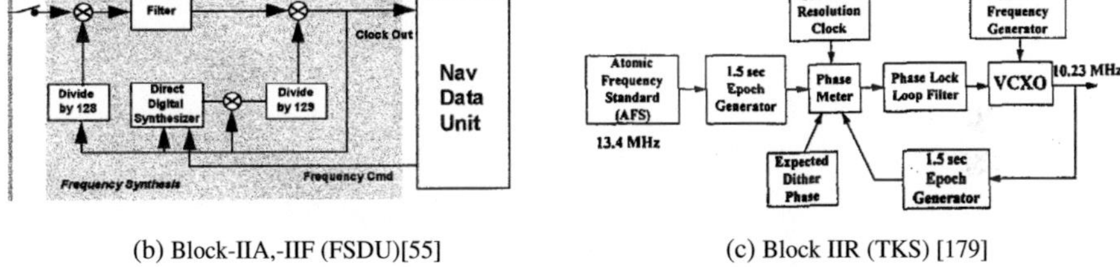

(b) Block-IIA,-IIF (FSDU)[55]

(c) Block IIR (TKS) [179]

Fig. 4.5: GPS frequency distribution units

The Block IIR TKS unit deserves further attention, as this unit modifies the noise characteristics of the AFS signal. A detailed description of the TKS architecture can be found in [11, 138]. The unit uses two frequency sources. One is an atomic frequency source and the second a Voltage-Controlled Crystal Oscillator (VCXO) at 10.23MHz locked by a Phase-Lock-Loop (PLL) to the AFS. The VCXO provides the navigation frequency. The phase difference between the two signals is precisely measured by a hardware phase meter. The output of the phase meter is the driving signal for the TKS PLL which is implemented in the software. This software also implements the Selective Availability capabilities to intentionally degrade the clock signal to non-authorized users. A phase meter predictor uses clock models from the AFS and VCXO to predict the phase. The predicted phase is compared to the measured phase difference. Following this prediction and measurement, the actual phase of the VCXO is steered in order to correct for frequency drift, temperature sensitivity, phase and possible frequency steps of the AFS. The Allan deviation for each element and final TKS noise is provided in [179] and further complemented in [180] with the final noise figures:

$$
\begin{array}{ll}
\text{VCXO} & 3.1 \times 10^{-14}\tau^{-1/2} + 1.0 \times 10^{-12} \\
\text{AFS(RAFS)} & 5.0 \times 10^{-12}\tau^{-1/2} + 4.2 \times 10^{-14} \\
\text{Phase Meter} & 1.7 \times 10^{-10}\tau^{-1}
\end{array}
$$

There are three major noise sources in the TKS system: the AFS, the VCXO, and the phase meter. The VCXO has lower noise at short time intervals than the AFS. The phase meter noise is higher, but with a higher slope (τ^{-1}). The bandwidth of the PLL is chosen to achieve the best Allan Variance in order to be driven by the VCXO at short term and the AFS at long term. In orbit, paradoxically, the automatic detection of anomalies caused several anomalies first reported by [36]. Further analysis by [180] pointed to the thermal sensitivity of the VCXO during eclipse phases forcing to decrease the PLL integration time and increasing the short term noise. The final integration time seems to be the same for all Block-IIR satellites as later observed in Figure 6.21 where all satellites have the same noise transition from the VCXO to the AFS.

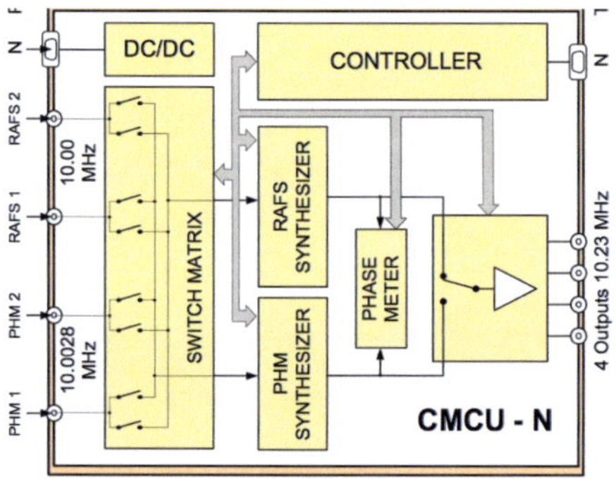

Fig. 4.6: Galileo CMCU design [53]

The frequency distribution unit on-board Galileo satellites is called CMCU (Clock Monitoring and Control Unit). GIOVE-A, -B and IOV share the same design developed by Thales [163]. A similar design was proposed by Astrium for FOC satellites [53]. Two clocks are powered on in Galileo satellites: one is used as the nominal clock for signal generation, while the other is kept switched on but not used for service provision. The CMCU allows the monitoring of the phase difference between redundant and nominal clocks phase through a phase meter. Due to the higher stability and the lower drift of the PHM, the phase meter measurements can be used directly to derive frequency drift and stability figures in order to characterize the rubidium clock. In case of the PHM failing the longer stabilization time of the RAFS would be already completed and the frequency stability verified, allowing for a quick reintroduction of the satellite into the constellation.

In the future, it will be possible to introduce even more complex systems. On ground, the reference time at timing laboratories is usually created with an ensemble of several AFS in order to have a continuous time scale. A dedicated algorithm processes the input of the different AFS in order to detect any clock anomalies, reject faulty units and introduce new clocks with the minimum impact in the time scale stability. The output of the algorithm is used to steer

GNSS	Block	FDU	Rb	Cs	PHM
GPS	I	DDS	3	1	
	IIA	FSDU	2	2	
	IIR	TKS	3		
	IIF	FSDU	2	1	
	III	TKS	3		1*
GLO	K	DDS		3	
	M	DDS		3	
Galileo	all	CMCU	2		2

Tab. 4.2: Timing subsystem

an active H-maser in order to create a physical realization of the time scale. Time ensemble algorithms are also used at USNO [101] and Galileo [156] to generate the system time. Adaptation of time ensemble algorithms applied on-ground for space applications would allow for autonomous detection of clock anomalies, and increase the reliability and integrity of the timing subsystem. Solutions have been proposed for military satellite communications (milsatcom) systems [34], as enhancements to GPS TKS [141] and for Galileo [54]. To keep several AFS active requires good isolation, more power consumption and energy dissipation capabilities demanding larger satellite dimensions. In the short future a single robust clock seems to be the most convenient and simplest strategy for navigation satellites (as demonstrates the removal of the TKS in GPS Block IIF). Nevertheless, plans exist for GPS-III to have a time keeping system similar to Galileo in order to operate and monitor a backup experimental AFS for stability performance measurements and characterization [175].

4.4 Payload delays

The presence of different group delays between signals is a well-known feature. It was acknowledged by system design with the inclusion of a dedicated group delay correction in each GNSS navigation message. GPS used manufacturer calibration biases until these were replaced by ground estimations from 1999 on [176]. From 2000, IGS has also considered delays between different modulations of the same frequency C1-P1 for the generation of the final products (see IGSMAIL-2827). However, the increasing number of modulations in GNSS is highlighting the importance of these delays when combining different signals or modulations inside the same system [162] or when combining different systems [70].

The previous two sections have described the timing subsystem. However, the payload is composed of several units before the signal is finally radiated to the user. Hereafter, the path followed by the navigation signals will be briefly explained based on the GIOVE payload, as shown in Figure 4.7 including the differences between GIOVE-A and -B manufacturer units. Further details on GIOVE satellites can be found in the early design [21], the final satellite

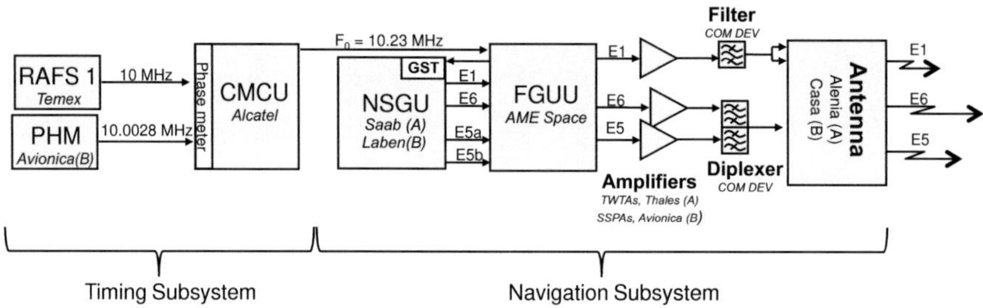

Fig. 4.7: GIOVE payload

manufacturer description [147, 96] or some independent summary of overall publications as collected in [87].

The satellite performs several steps before it radiates the navigation signal. Firstly, the atomic frequency standard generates the frequency signal close to 10 Mhz. The frequency distribution unit selects the active clock and converts the frequency to the basic frequency (10.23 MHz).

Secondly, this reference signal is used by the Frequency Generator and Up-converter Unit (FGUU) to generate the stable carrier provided to the navigation unit. Based on this signal the Navigation Signal Generation Unit (NSGU) generates the local pulse per second signal used to tick the satellite time. The navigation data and spreading codes are formatted, encoded and modulated using this LPPS signal and the satellite time then inserted into the navigation message time-tags. Next the signal is passed again to the FGUU for up conversion (e.g. L1 = 10.23MHz $\times$1540). Three separate amplifiers (SSPAs or TWTA) amplify the modulated navigation signals. An output multiplexer is required to combine the output signals from the two amplifiers of the low band channel signals (E5a+E5b and E6) with the antenna low band input. The primary function of the filter is to define the high band channel (L1) before passing it to the antenna. High rejection of spurious signals outside the navigation bands is also provided by both the filter and multiplexer.

Thirdly, the L-Band antenna takes the filter and multiplexer outputs and transmits the signal to the ground using right-handed circular polarization. Further frequency dependent variations are introduced by the navigation antenna. The phase as radiated by the antenna will present deviations with respect to a perfect sphere.

Once the signal is modulated, the different signals follow different circuit paths untill they are finally radiated to the user by the L-Band antenna. Group and phase delays of the signals from the navigation unit to the antenna phase center are different. Thermal variations and thermal sensitivity are different for each unit; as a consequence, any temperature change will affect each signal in a different way. These delays have to be assumed as constant with variations at the nanosecond level mainly due to temperature changes. Any change in the payload or receiver chain will affect the absolute value however.

Finally, in the ground reception chain, depending on the signal modulation the asymmetry of the signal (the so called S-curve bias) generates an additional delay in the receiver. In the case of GLONASS, due to the frequency division multiple access (FDMA) applied in the signal transmission by each satellite, the frequency dependency on the receiving chain makes the payload delays different for each satellite. Consequently, the bias cannot be easily absorbed by the receiver ionosphere-free clock.

In summary, the overall frequency dependent delay that is present in a GNSS measurement can be attributed to a combination of four distinct sources [31]:

1. the propagation delay through the transmit/receive chain;

2. the phase delay, dependent on the phase response of the transmit/receive chain;

3. the group delay, dependent on the amplitude and phase response of the transmit/receive chain;

4. the asymmetry of the correlation function, dependent on the amplitude and phase response of the receive chain.

4.5 Navigation antenna delays

The bias introduced by the navigation antenna deserves dedicated attention as it is the main source of phase and group delay variations with respect to the geometry. Figure 4.8 introduces the phase center variations in the navigation antenna for a single frequency. The location of the navigation antenna reference point (ARP) and laser retro reflector (LRR) position with respect to the satellite reference frame (SRF) is known by design, calibrated by the manufacturer with sub-millimeter accuracy and common to all frequencies. Orbits and clocks are, however, computed with respect to the satellite center of mass (CoM). The mass and center of mass of the satellite are important for the satellite integration with the launcher and as a result its position is accurately calculated by the satellite manufacturer. The dry mass is calibrated after satellite integration and refined after the satellite is filled with the final propellant before the launch. The main difficulties arise in the calibration of the center of phase of the antenna from where the signals are radiated.

The antenna reference point is further displaced to the so-called Center of Phase (CoP) which can be seen as the center of the sphere minimizing the phase center variation (PCV) to this point (see ANTEX 1.3 definition). The center of phase and variations can be obtained from

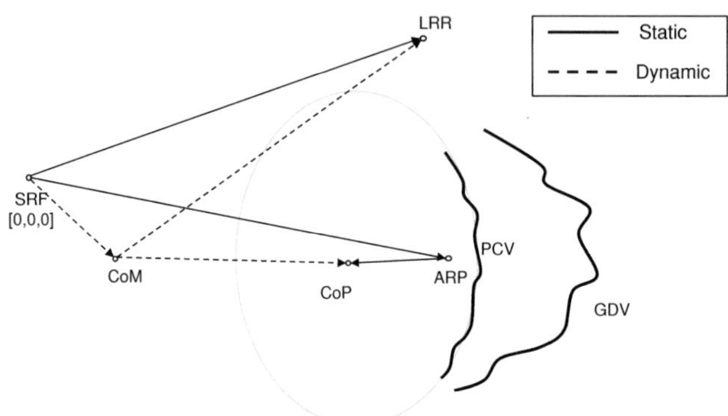

Fig. 4.8: Navigation representative centers

the navigation antenna calibrations, or be estimated by precise orbit determination with some relative uncertainty due to the correlation with other estimated parameters. Normally these variations account for up to several millimeters in phase (i.e. rather below 10 picosecond at single frequency).

Group delay variations (GDV) associated to the antenna due to the navigation antenna amount up to several decimeters at a single frequency and have a different center location [126]. Dependency on the transmitted modulation and receiver configuration parameters (bandwidth and correlator spacing) as highlighted by [70] limits the possibility of correcting for this delay in an effective form. They are usually ignored by the user community due to the difficulties in its estimation and correction and are considered an additional noise to be averaged. However, in the future their contribution should be considered more throughly as they could be partially responsible for the boundary jumps observed between different day solutions.

4.6 Conclusions

In this chapter, the physical elements in the satellite in charge of generating the timing signal have been presented. First, the AFS generates a reference frequency at f_i. Second, a frequency synthesizer in the timing subsystem converts this frequency to the reference frequency F_0 as observed on ground. Then the satellite time scale ('real clock') is created by counting AFS cycles (10.23E6 cycles = 1 second) in the navigation signal generation unit. Finally, the 'Timing Signal' is created by the navigation unit when encoding the navigation codes over the phase provided by the timing subsystem.

In Section 4.2, the AFS have been reviewed by technology type from the early years to the current state of the art and in the light of future trends. Performance metrics have steadily increased over the last 30 years, beginning with the first dedicated RAFS in GPS adapted from

German ground technology. Nowadays, new PHM in Galileo and RAFS in GPS have already yielded new possibilities for navigation and POD. Also, new technologies have appeared in the optical domain which have dramatically improved the current AFS performance. Nevertheless, these technologies still need to become mature and to meet space requirements before being embarked in a GNSS satellite. All these items point to the fact that a mixture of different AFS are being used in-orbit, but any application should take into account the diversity and particularities of each AFS in each GNSS.

The frequency distribution unit (FDU) is an important element of the timing subsystem as it allows a user a switch between the different AFS on-board and an adjustment of their frequency. AFS noise characteristics can be modified in this step. In contrast, as a-priori satellite clock performance information, the AFS specifications are broadly used. Section 4.3 highlights how this hypothesis can lead to erroneous conclusions. The timing subsystem is composed of two elements: the AFS and the frequency distribution unit. This is particularly true for Block-IIR satellites where the timing signals performance shall be taken from the combination of the AFS(rubidium) plus the settings of the Time Keeping System (TKS), which significantly increase the short term noise of the output signal. The frequency distribution unit can also implement the autonomous detection of any clock anomalies, and subsequently increase the reliability and integrity of the timing subsystem. This approach was implemented in the TKS and is being reviewed for future Galileo and GPS-III satellites. However, in the short future, a single robust AFS seems to be the most convenient and simplest strategy for navigation satellites; as has been demonstrated by the reintroduction of a similar design from Block IIA in GPS Block IIF for such unit.

Group delays particularly affect GLONASS receivers. A GNSS receiver identifies each GLONASS satellite by its unique frequency allocation, while recognizes other GNSS satellites by a common frequency allocation per system and a different code allocation per satellite. This frequency division implies that the group delays are satellite dependent and, therefore, not fully absorbed by the station clock as a common error for all satellites but by the error budget. Finally, Section 4.4 presents how frequency dependent phases and group delays are introduced in the timing signal broadcast to the user by the additional elements of the satellite navigation chain. These group delays have a constant part plus a daily variation associated to temperature fluctuations during the orbit period. The same principles are applicable to ground receiver chains. Depending on the signal modulation, the asymmetry of the broadcast signal generates an additional delay in the receiver (geometry dependent) which further limits the accuracy of time transfer. This effect is currently not accounted for by any processing.

5 Methodology applied in geodetic time transfer

5.1 Introduction

The estimation of the offset between the system and satellite time dt^s is performed by POD adjustment software where the true distance between the satellite and receiver is computed together with other parameters. Code measurements have been introduced as an absolute one-way time transfer from the satellite to the receiver. However the raw measurements do not provide the pure distance between the satellite and the receiver. The code measurement is also often called pseudorange as it includes the true range from the satellite to the receiver plus additional delays caused during the generation, propagation, reception and measurement of the signal. Several strategies exist to overcome the additional delays. They can be modelled with high accuracy (such as relativity), cancelled by linear combinations of different measurements in separate frequencies (as ionosphere), estimated (as clocks), lumped with other terms (as group delays) or simply averaged (as multipath). The clock offset represents only one of these additional delays and is traditionally considered by the geodetic community as a by-product of the satellite and station coordinates estimation.

The basic input data of POD estimations are receiver code and phase measurements. GIOVE satellites allow for free tracking of 7 different modulations (E1A, E1B, E5a, E5b, E5, E6A and E6B) on four separated frequencies. These modulations can carry different information in phase or in quadrature leading to a significant number of tracking configuration possibilities in the receiver with different associated hardware delays. In the standard format Rinex 3.00 a total of 18 different types of tracking codes are allocated to Galileo and 14+2 to GPS, this number being even further increased in version 3.01. Still, despite the numerous modulations and frequencies available only two single frequency measurements are used in POD . Additional measurements are normally ignored and all solutions referred to a basic ionosphere-free combination even by the GNSS service provider (e.g. E1B-E5a for Galileo open service).

The combination of two single frequency measurements into a single quantity as a basic input observation in POD carries some consequences in the timing area. Signals are not aligned at the output of the satellite antenna as explained in Section 4.4 nor in the reception chain, resulting in different group delays associated to each signal. The group delay between the signals gets lumped into the estimated 'ionosphere-free clock' and consequently the real clock offset is no longer estimated. Furthermore, if two different GNSS constellations with different basic pairs of signals are mixed, the 'ionosphere free clock' at the station becomes GNSS dependent and an additional inter-system bias (ISB) needs to be computed. Not all applications rely on dual frequency combinations. Some applications require the estimation of the group delays, such as ionosphere estimations for space weather applications or single frequency users, with the majority of GPS users relying only on GPS C/A code whereas the message refers to P1-P2 combination. Such group delay estimations can provide a way to retrieve the difference between the 'ionosphere-free clock' and the real signal clock.

This chapter reviews the state of the art of the clock estimation. In the current approaches the group delays are lumped into the ionosphere-free clock and considered constant for each day. This hypothesis is often ignored and conditions the accuracy of the estimated clocks and any conclusion derived from these estimations. As a consequence, special attention will be given to the group delays. First, methods are briefly reviewed together with IGS product combinations normally used as benchmark. Second, the ionosphere-free combination is revised and the other parameters included in the estimated 'ionosphere-free clock' are identified. From these parameters the group delay is identified as the main bias. Third, the estimation of group delays together with ionosphere estimations is also reviewed. Finally, a practical example of group delay and inter-system bias estimation is presented in Section 5.5 with GIOVE satellites using standard and novel methodologies.

5.2 Time transfer in metrology

National time laboratories are in charge of creating a continuous realization of UTC(k) based on their more reliable atomic frequency standards while maintaining traceability to the international realization of UTC. Additionally, the laboratories may be involved in the realization of the next generation of atomic clocks which need to be compared against other frequency standards at other laboratories with comparable stability. Time laboratories need to be compared with the best available time transfer techniques to create TAI, whereas new picosecond techniques are currently under development aiming at frequency transfer at 1E-16 and beyond in order to allow time transfer of uprising optical clocks.

Some of these laboratories are in charge of maintaining the legal time for their country and for its distribution to users. Radio clocks synchronized to terrestrial time signals are still broadly used to distribute time to mass market users. For over fifty years the Physikalisch-Technische Bundesanstalt (PTB) has disseminated time signals by means of the low-frequency transmitter DCF77. In addition, since the 1970s legal time for Germany has been broadcast using coded time information by way of amplitude modulated second markers. This link is the most important medium for the dissemination of legal time by PTB.

Other radio and time signal emissions exist in Europe broadcasting UTC(k), in accordance with the recommendation 460-4 of the International Telecommunication Union (ITU), namely EBC in Spain, HBG in Switzerland, MSF in the United Kingdom and TDF in France. Some of these other signals are not broadly used, such as HBG discontinued in 2011 or EBC with limited range and out of service in 2012, or TDF due to the more expensive equipment required to use this signal. Instead, the radio signal DCF77 transmitted by PTB is received widely across continental Europe and represents the most used time signal on the continent.

Today, approximately half of all 'large electrical clocks' (table clocks, mounted clocks, wall clocks and alarm clocks) sold in the private sector are radio-controlled clocks. In addition, more than half a million radio-controlled industrial clocks are in use. Such receivers are implemented in railway, air-traffic control, parking meters, traffic lights, heating and ventilation systems, telecommunication, energy-supply industries, time-related tarifs for billing services, radio and television stations, and network time servers which feed the time received from DCF77 into computer networks and further distribute it to other users, such as stock markets [132].

The development of the internet moved the open air radio signal distribution to the wire as the major mean for transmitting time. The Network Time Protocol (NTP), based on the Internet IP protocol, aims at synchronizing computer clocks within LANs and throughout the Internet with accuracy to the millisecond. NTP requires time servers which distribute their time to other computers. While most time laboratories provide this service - as an example PTB operates three time servers - many of the servers use direct access to GNSS time.

Since the year 2000, GPS devices have rapidly expanded to mass market users through personal digital assistants (PDAs) and dedicated navigation devices such as TomTom and Garmin. Furthermore, mobile phones are by far the most pervasive consumer electronics devices globally. Emergency call mandates for reliable position and smartphone expansion have decreased the price of GNSS receiver technology and extended the availability of accurate GNSS time to a large number of users.

All of these activities from specialized time laboratory equipment to mass market users require the transmission of time by one-way or two-way techniques explained hereafter.

5.2.1 One-way time transfer

The most common way to transfer time is through a transmitted signal, usually an electromagnetic (radio) wave. However, there are several variations on this approach. In one-way time transfer, the source, A, sends a time signal to the user, B, through a transmission medium with a delay, d, over a transmission path. Therefore, some correction for this delay is required unless the accuracy requirement is very relaxed, as it is the case for the popular radio-controlled clocks based on long wave and shortwave transmissions.

GNSS time transfer is also a clear example of one-way time transfer. As explained in Chapter 3.8, each GNSS satellite generates a time signal traceable to an official realization of UTC(k) on-ground and each code measurement represents an absolute one-way time transfer between the satellite and the receiver. If high accuracy time transfer is desired in a one-way system the physical locations (coordinates) of the two clocks must be known so that the path delay can be calculated. This is the case for GNSS receivers where the coordinates can be computed in advance for static users or together with the time offset for dynamic users.

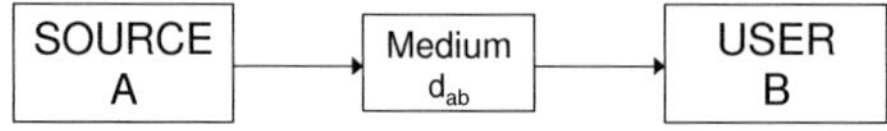

Fig. 5.1: One-way time transfer.Source: NIST website

From the early 80's, GPS time transfer started to be used for time distribution between time laboratories using common view technique, in order to eliminate ephemerides and Selected Availability errors, with an accuracy of 10 ns (rms) while one-way was still limited to 100 ns [4]. After the deactivation of Selected Availability on 2 May 2000, one-way GPS time transfer became accurately available as a global common navigation time reference with an accuracy of 6.32 ns (rms) [58]. Nevertheless, the timing community still relayed in common view due to its simplicity.

In the common view technique, two stations, A and B, receive a GNSS signal simultaneously from a single transmitter and measure the time difference between this received signal and their own local clock as depicted in Figure 5.2

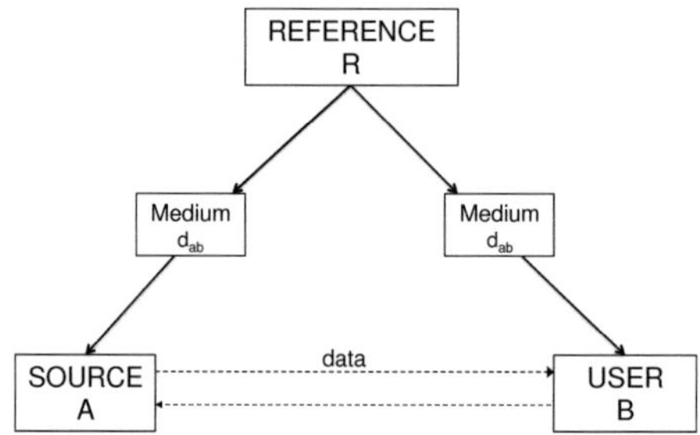

Fig. 5.2: Common view time transfer. Source: NIST website

The data are then exchanged between receivers A and B using any convenient method. Common offsets, such as satellite orbit and clock errors cancel in the differentiation. Receivers have to be close enough to reduce thepropagation delay differences between both lines of sight. Only a differential calibration is required between the two receiving stations. A differential calibration is made to a BIPM traveling receiver temporally located at the station.

The GPS common-view technique has been used for many years by the Bureau International des Poids et Mesures (BIPM) as one of its main techniques for international time comparisons. Only code measurements are typically used in this solution, neglecting the two orders of magnitude more precise carrier phase measurement widely use in POD.

Many of the geodetic GNSS receivers hosted in national timing laboratories operate continuously within the International GNSS Service (IGS) and their clock offsets are estimated by geodetic techniques with sub-nanosecond accuracy. This was recognized when the International GNSS Service (IGS) and the Bureau International des Poids et Mesures (BIPM), formed a joint pilot study to analyze IGS Analysis Centers clock solutions and recommended new means of combining them.

Only a few time laboratories hosting GNSS dual frequency receivers are included in the IGS network. Due to the simplicity, low cost and automation possibilities of PPP, its use for time transfer is increasing in the time community. The PPP method is a post-processing approach using undifferenced observations collected from a single geodetic GPS receiver along with satellite orbit and clock products. Parameters estimated in PPP are station positions (in static or kinematic mode), station clock states, local troposphere zenithal delays and carrier phase ambiguities. The best position solution accuracies, reaching a few centimetres in horizontal coordinates and less than 10 cm in vertical coordinates (RMS), are obtained by processing GPS dual-frequency pseudorange and carrier phase observations with IGS precise satellite orbit and clock products.

Results obtained by POD and PPP are of similar quality [33]. PPP is used to compute TAI time links since 2009 with 30 participating time laboratories (7-8 also IGS stations) and rapid orbits. Finally, geodetic time transfer is slowly becoming one of the official three methods for time transfer in the BIPM.

5.2.2 Two-way time transfer

Two-way time transfer involves signals that travel both ways between the two clocks or oscillators that are being compared, as shown in Figure 5.3. The measurements between both stations are differentiated and the delay due to the propagation medium is cancelled. Internet time transfers using the Network Time Protocol (NTP) is an example of this technique.

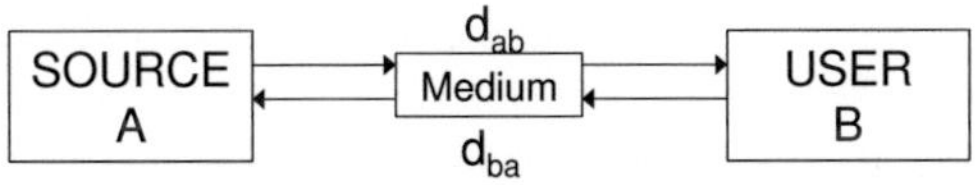

Fig. 5.3: Two way time transfer. Source: NIST website

The Two-Way Satellite Time and Frequency Transfer (TWSTFT) technique provides stable and accurate time transfer since nearly all of the propagation delay cancels out due to symmetry. A significant disadvantage of TWSTFT is the added complexity due to the need for both transmitter and receiver hardware at each station and a dedicated synchronization of the measurements to the telecommunication satellite. It is also more expensive, since it requires paying for satellite time if a commercial communication satellite is used. More precise than GPS common view

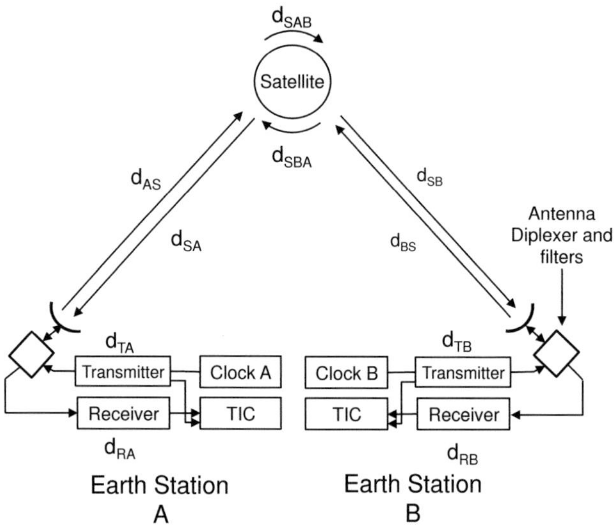

Fig. 5.4: Two-way satellite time and frequency transfer (TWSTFT) [66]

from its introduction in 1989, it represents one of the main methods for time transfer between time laboratories [66].

SLR based time transfer is a new two-way technique which requires that laser stations and satellites be equipped with a photo-detection system, a time-tagging device and a retro-reflecting device. The innovation in this technique is the introduction of a sensor on-board to detect the laser beams. It then becomes possible to deduce the time-transfer between stations A and B from the difference between d_{AS} and d_{BS}. T2L2 is an experimental time transfer by laser link embarked as a passenger instrument in the Jason-2 satellite launched on 20 June 2008. The T2L2 uses also a two-way technique based on satellite laser ranging with a target of 1 ps time transfer precision and a level of accuracy better than 100 ps depending on the calibration of the SLR station. Mission objectives and principles are available at http://smsc.cnes.fr/T2L2.

The ACES clock signal will also be transferred to ground by a two-way time and frequency transfer link in the microwave domain. The Atomic Clock Ensemble in Space (ACES) is an ESA mission in fundamental physics based on the performance of a new generation of atomic clocks (PHARAO and SHM) to be embarked in the International Space Station (ISS). These clocks have been briefly explained in Section 4.2 as future AFS. Time transfer between distant time laboratories will be possible by common view to ACES by using a new frequency transfer link [30]. Optical fiber time transfer through a fiber network is a promising new two-way technique which is used for short experimental links but which is improving steadily in range, cost and accuracy.

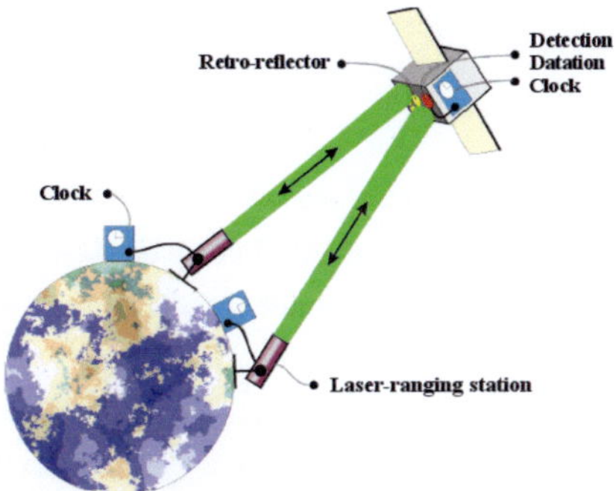

Fig. 5.5: Principle behind SLR based time transfer. Source: CNES

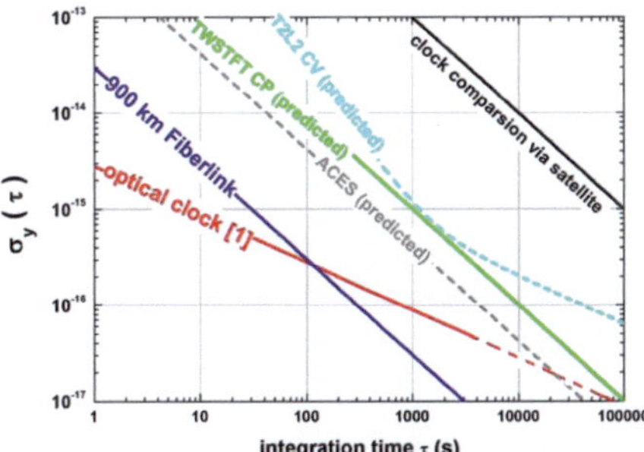

Fig. 5.6: New time transfer techniques and optical clocks stability. Source: PTB

These three new techniques (T2L2,ACES and fiber) are under experimentation by some time laboratories, such as PTB in charge of generating UTC(Germany) [133], and they could be operationally available in the future to complement GNSS and TWSTFT. Currently, it should be noted that GPS(PPP) and TWSTFT, although used at only 30% of time laboratories, account for between 70-75% of the clock weight in TAI [130].

5.3 Orbit and clock determination

5.3.1 Methodology

Precise orbit determination (POD) software packages are in charge of solving for orbits, clocks and other model parameters using a priori estimates and a weighted least squares estimation by a differential correction to the receiver observations. Several software package and strategies exist to estimate orbit and clocks from the information provided into the analysis strategy summary files (*.acn) of each IGS analysis center, which may may be separated into two main adjustment methodologies for calculating a dynamic orbit by:

- Batch weighted least squares adjustment

- Kalman filter

Batch based dynamic algorithms, using very precise dynamic and measurement models are the preferred approach. The batch weighted least squares method is currently the preferred solution among IGS analysis centers (13/15) and the Galileo ground segment whereas the Kalman filter is used by a subset of analysis centers (2/15) and the GPS ground segment.

A second important strategy is the selection of the basic modeled observable. Here also two basic strategies can be identified depending on the basic observable used in the adjustment.

- Double differenced

- Un-differenced

The first approach consists in the use of double difference observations between two satellites and two stations at the same epoch. In the new observation the clock corrections, hardware delays for receiver and satellite as well as the initial phase shift term cancel out or may be neglected, giving access to the integer nature of the initial phase ambiguities. The combination of four different measurements into one observation may also increase the correlation between the parameters and the noise level [35, p. 146] . The advantage is the possibility of using ambiguity resolution constraints and the easier pre-processing. Nevertheless, the clock parameters require a second estimation step with the code measurements where the orbit solutions are fixed to the previous estimates. It has to be remarked, that if static baselines are formed (such as in Bernese [35]) a larger number of tracking stations is required.

The second approach is the use of un-differenced measurements, where the clock parameters and orbits are obtained in a common adjustment. Both strategies are equally used at IGS, where half of the analysis centers implemented solution or the other.

Orbit and clock solutions are strongly correlated in both cases, especially in terms of the radial orbit component. The combination of different clock estimates obtained by different software packages, data or estimation strategies requires an additional radial correction to the combined orbit, as well as an alignment of the time scales [86].

5.3.2 The weighted batch least squares adjustment

No precise orbit determination software or adjustment strategy seems to provide a clear advantage over the other ones. The strong point of IGS combination products is the freedom of different strategy selection by each analysis center and the quality control by comparing each solution against a weighted combination. All adjustment strategies share similar principles: given an initial state vector (for the dynamics and time) at the epoch t_0, the orbit propagation computes the satellite position at the receiver measurement times over an integration arc using a model of forces acting on the satellite. Then, a delta correction to the initial state vector is estimated in order to minimize the residual to the modelled observable. In order to understand the correlation between the observations and estimated parameters, the basic principle of un-differenced batch weighted least squares solution will now be reviewed. The principle is explained in numerous textbooks and publications, such as [178, 25, 116] used for the derivation of the mathematical and stochastic model performed hereafter.

Mathematical model fundamentals

Given a number M of measurements l dependent on N unknown variables x the over-determined system of linearized observation equations $M > N$ can be constructed as :

$$\begin{bmatrix} l_1 \\ l_2 \\ \vdots \\ l_M \end{bmatrix} + \begin{bmatrix} v_1 \\ v_2 \\ \vdots \\ v_M \end{bmatrix} = \begin{bmatrix} F_1(x) \\ F_2(x) \\ \vdots \\ F_M(x) \end{bmatrix} \tag{5.1}$$

which represents the Gauss-Markov model:

$$l + v = \bar{l}(x) \tag{5.2}$$

where $\bar{l}(x) = F(x)$ denotes the functional relationship between the observables and the unknown parameters. Normally, this relationship is not linear and the system needs to be linearized by a Taylor expansion of F around a chosen initial value x_0 where only the first term is retained

$$\bar{l}(x) \approx F(x_0) + F'(x)(x - x_0) = F(x_0) + F'(x)\Delta x_0 \tag{5.3}$$

and the single prime $'$ denotes the first order derivative. Since x has N parameters $x = [x_1, x_2, \ldots, x_N]$, by including the partial derivatives the relation can be rewritten as:

$$\bar{l}(x) \approx F(x_{1_0}, \ldots, x_{N_0}) + \frac{\Delta F}{\partial x_1}\Delta x_{1_0} + \ldots + \frac{\Delta F}{\partial x_N}\Delta x_{N_0} \tag{5.4}$$

The matrix A can now be constructed with all the partial derivates,

$$A = \frac{\Delta F}{\partial x} = \begin{bmatrix} \frac{\Delta F_1}{\partial x_1} & \cdots & \frac{\Delta F_1}{\partial x_N} \\ \vdots & \ddots & \vdots \\ \frac{\Delta F_M}{\partial x_1} & \cdots & \frac{\Delta F_M}{\partial x_N} \end{bmatrix} \tag{5.5}$$

where each partial derivate is evaluated at x_0, the relation can be expressed as

$$\bar{l}(x) \approx F(x_0) + A\Delta x_0 \qquad\qquad [5.6]$$

Finally, to simplify the subsequent derivations the nomenclature is changed by using $=$ instead of $\approx$, $\Delta x_0 = x$ and $F(x_0) = l_0$:

$$\bar{l}(x) = l_0 + Ax \qquad\qquad [5.7]$$

In general, the desired solution of this over-determined (M>N) system of equations is the so-called L2-Norm solution which minimizes the sum of the squares of the measurement deviation or residuals v with respect to the model $\bar{l}(x)$:

$$\Phi(x) = \sum p_i v_i^2 = v^t P v \qquad\qquad [5.8]$$

where P is the symmetric matrix of observation weights p_i. This relation can also be re-written for each equation as:

$$\Phi(x) = v^t P v = \bar{l}(x)^t P \bar{l}(x) - 2l^t P \bar{l}(x) + l^t P l \qquad\qquad [5.9]$$

The minimum is obtained by a partial derivation $\frac{\partial \Phi(x)}{\partial x} = 0$,

$$\frac{\partial (v^t P v)}{\partial x} = 2\bar{l}(x)^t P \frac{\partial \bar{l}(x)}{\partial x} - 2l^t P \frac{\partial \bar{l}(x)}{\partial x} + 0 = \left(\frac{\partial \bar{l}(x)}{\partial x} \right)^t P \left(\bar{l}(x) - l \right) = 0 \qquad [5.10]$$

by sustitution of 5.7 into 5.10:

$$A^t P A x = A^t P (l - l_0) \qquad\qquad [5.11]$$

the final solution vector of unknowns is obtained as :

$$\hat{x} = \left(A^t P A \right)^{-1} A^t P (l - l_0) \qquad\qquad [5.12]$$

where $N = A^t P A$ is the normal equation matrix which must be a regular matrix in order to be invertible and $\hat{x}$ represents the delta (Δx_0) to the initial vector x_0. Since A and l_0 depend on this initial state vector, in general, an iterative solution is required. Starting from a given state vector x_0, at each iterative step i an updated solution is achieved by evaluation of $A(x_i)$ and l_i, and subsequent resolution of $\hat{x}_{i+1} = \hat{x}_i + \Delta\hat{x}_i$.

Stochastic Model

The more precise an observation, the higher shall be the weight and as a consequence, the smaller the variance the higher is the weight applied. The matrix P contains the weights which are inversely proportional to the variances and multiplied by the a priori variance σ_0^2 as scale factor:

$$
P = Q_{ll}^{-1} = \sigma_0^2 C_{ll}^{-1} =
\begin{bmatrix}
\frac{\sigma_0^2}{\sigma_{l_1}^2} & 0 & \cdots & 0 \\
0 & \frac{\sigma_0^2}{\sigma_{l_2}^2} & \cdots & 0 \\
\vdots & \vdots & \ddots & \vdots \\
0 & 0 & \cdots & \frac{\sigma_0^2}{\sigma_{l_M}^2}
\end{bmatrix}^{-1}
\tag{5.13}
$$

From the law of error propagation applied to Equation 5.12 the cofactor matrix of the unknowns $Q_{\hat{x}\hat{x}}$ is obtained as :

$$
Q_{\hat{x}\hat{x}} = F Q_{ll} F^t = (A^t P A)^{-1},
\tag{5.14}
$$

as well as the cofactor matrix of the adjusted observations $Q_{\hat{l}\hat{l}}$

$$
Q_{\hat{l}\hat{l}} = A^t Q_{xx} A,
\tag{5.15}
$$

and the cofactor matrix of the residuals.

$$
Q_{vv} = Q_{ll} - Q_{\hat{l}\hat{l}}.
\tag{5.16}
$$

The covariance matrices are finally obtained by multiplication with the a posteriori variance $\hat{\sigma}_0^2$.

$$
C = \hat{\sigma}_0^2 Q
\tag{5.17}
$$

where

$$
\hat{\sigma}_0^2 = \frac{v^t P v}{M - N}
\tag{5.18}
$$

fulfills the global test:

$$T_G = \frac{\hat{\sigma}_0^2}{\sigma_0^2} \sim F_{r,\infty} \qquad [5.19]$$

if the applied Gauss-Markov model is true.

Application to POD

The described functional mathematical and stochastic models can now be applied to the orbit and clock computation process based on the following scheme:

1. Given an initial satellite state vector x_0, an orbit is produced by numerical integration of the equations of motion of the satellite over the estimation period. Initial clock offsets to the reference time scale are also computed for satellite and stations.

2. Using the predicted information and known station positions, the distance between station and satellites is computed and the vector $\bar{l}(x_0)$ created.

3. The differences between the observations l and predictions $\bar{l}(x_0)$ are computed.

4. The matrix $A = \frac{\partial \bar{l}(x_0)}{\partial x}$ is created.

5. The weight matrix P is formed as a diagonal matrix based on the a priori variance assigned to the observations l.

6. The sum of squares of the residuals is minimised by estimating the dx corrections to the initial orbit, clocks and other parameters.

7. The a posteriori standard deviation $\hat{\sigma}_0^2$ is computed. The tolerance for ending the iterative computation is checked and the process is repeated until the desired convergence is achieved.

Let us briefly consider each step. First, an initial state vector position is required. X_0 for orbit and clock states can be extracted from the propagation of previous estimations, satellite transmitted navigation message, public orbital two-line elements generated by North American Aerospace Defence Command (NORAD) or on S-band tracking (e.g. Galileo orbits routinely provided to ILRS for satellite ranging are based on S-band). The initial position vector needs to be propagated. The solution of the problem is achieved by numerical integration of the equations of motion, which can be expressed in matrix form as follows:

$$X = \begin{bmatrix} \vec{r} \\ \vec{v} \end{bmatrix} = \int_{t_0}^{t} \dot{X} dt + X_0 \qquad [5.20]$$

where the principal forces acting on the satellite are due to the Earth's gravitational potential (complemented by tidal forces), perturbation due to the gravitation force of third bodies, perturbing accelerations due to solar radiation pressure and other accelerations which may affect the satellite, as manoeuvres.

$$\frac{dX}{dt} = \dot{X}_{Earth} + \dot{X}_{3^{rd}-body} + \dot{X}_{SRP} + \dot{X}_{other} \qquad [5.21]$$

The complete initial state vector depends on the model and the estimation strategy applied but normally is composed of a first part with non-clock parameters related to the geometry and signal propagation, and a second larger part containing the clock offsets. The first part contains the satellite initial position $\vec{r}^s$ and velocity $\vec{v}^s$ (6 per arc), the satellite solar radiation pressure SRP^s parameters (usually 5-9 radiation pressure parameters are estimated, as explained in Section 7.4.3), station troposphere zenith delay T_r (normally 24 per day per station), ambiguities N_r^s (minimum 1 per pass per station), and Earth orientation parameters (3 per arc). The second part includes the satellite dt^s and receiver dt_r clock offsets for each epoch (amounting to $N_{epochs} \times (N^{sats} + N_{rec})$), and the receiver inter-system biases ISB_r in the case of combined GNSS processing (1 per station).

$$X_0 = \begin{bmatrix} \vec{r}^s \\ \vec{v}^s \\ SRP^s \\ T_r \\ \lambda N_r^s \\ ERP \\ ISB_r \\ dt_r \\ dt^s \end{bmatrix}_{(N \times 1)} \qquad [5.22]$$

Second, the reconstructed or expected observation from the receiver to the satellite is constructed as the geometrical difference between both positions. Station coordinates (x_r, y_r, z_r) are fixed from previous estimations (e.g. ITRF),

$$\rho_{X_0} = \sqrt{(x_0^s - x_r) + (y_0^s - y_r) + (z_0^s - z_r)} \qquad [5.23]$$

and the residual vector dl is computed from the difference between the M observations between receiver r to satellite s and the propagation from initial state vector.

$$dl = l - l_0 = \begin{bmatrix} \rho_1 - \rho(X_0,t_1) \\ \rho_2 - \rho(X_0,t_2) \\ \vdots \\ \rho_M - \rho(X_0,t_M) \end{bmatrix}_{(M \times 1)} \qquad [5.24]$$

Third, the Jacobi matrix A which contains the partial derivatives of the computed observations with respect to the estimated parameters is numerically created:

$$A = \begin{bmatrix} \frac{\partial \rho_1}{\partial x_0} & \frac{\partial \rho_1}{\partial y_0} & \frac{\partial \rho_1}{\partial z_0} & \frac{\partial \rho_1}{\partial \dot{x}_0} & \frac{\partial \rho_1}{\partial \dot{y}_0} & \frac{\partial \rho_1}{\partial \dot{z}_0} & \frac{\partial \rho_1}{\partial \text{SRP}} & \cdots & \frac{\partial \rho_1}{\partial dt_r} \\ \frac{\partial \rho_2}{\partial x_0} & \frac{\partial \rho_2}{\partial y_0} & \frac{\partial \rho_2}{\partial z_0} & \frac{\partial \rho_2}{\partial \dot{x}_0} & \frac{\partial \rho_2}{\partial \dot{y}_0} & \frac{\partial \rho_2}{\partial \dot{z}_0} & \frac{\partial \rho_2}{\partial \text{SRP}} & \cdots & \frac{\partial \rho_2}{\partial dt_r} \\ \vdots & \vdots & \vdots & \vdots & \vdots & \vdots & \vdots & \ddots & \vdots \\ \frac{\partial \rho_M}{\partial x_0} & \frac{\partial \rho_M}{\partial y_0} & \frac{\partial \rho_M}{\partial z_0} & \frac{\partial \rho_M}{\partial \dot{x}_0} & \frac{\partial \rho_M}{\partial \dot{y}_0} & \frac{\partial \rho_M}{\partial \dot{z}_0} & \frac{\partial \rho_M}{\partial \text{SRP}} & \cdots & \frac{\partial \rho_M}{\partial dt_r} \end{bmatrix}_{(M \times N)} \qquad [5.25]$$

The weight matrix P is created by assigning a priori variances to the different measurements. Normally, code observations are weighted 100 times less than the phase. Since low elevation observations are affected from stronger multipath and measurement noise, as the signal power is lower due to the atmosphere attenuation, an additional elevation dependent weighting function $sin(e)^2$ is usually applied as a function of the elevation angle e. In the fourth step, the corrections to the initial state vector position are computed by resolving the matrix system in Equation 5.12, where $\hat{x}$ is the estimated correction to the initial parameters:

$$\hat{x} = (A^t PA)^{-1} A^t P(l - l_0)$$

For an IGS network type with up to 200 stations and 30 seconds sampling rate, the number of observations M is around 35 million per day considering only GPS satellites and ionosphere-free observations. To create the matrix A of dimension $M \times N$ and P as a diagonal matrix of dimension $M \times M$ imposes significant memory requirements. Instead, it is possible to directly create the smaller normal equation matrix $A^t PA$ (of dimension $N \times N$) and $A^t Pl$ (of dimension $N \times 1$) to save memory. The normal equation matrix depicted in Figure 5.7 is formed by all the non-clock parameters, the clock parameters and the correlations between them. As it is a symmetric matrix, it is also possible to store half of the data in order to further reduce the memory loss and processing time used in its construction and processing. The clock parameters are computed in snap shot by epoch. They represent about 80-95% of the unknowns to be computed depending on the size of the network.

Finally, since orbit determination is not linear, the process is repeated again. The recursion can be stopped by checking the a posteriori variance with respect to the expected distribution or by simply checking the gain in the correction dx at each step with respect to a threshold. A wise decision is to foresee a maximum number of iterations.

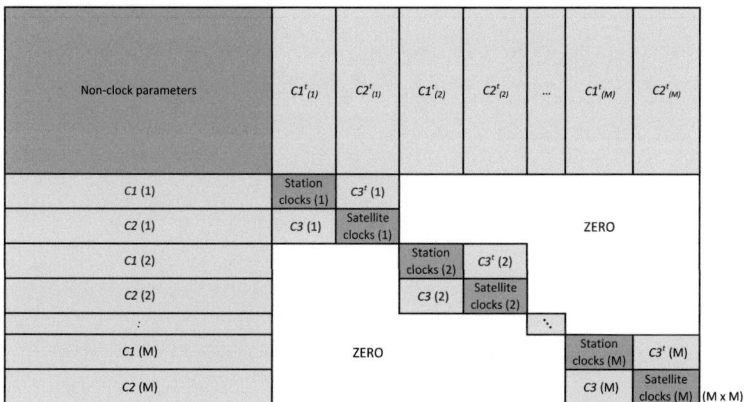

Fig. 5.7: Normal equation matrix structure

The weighting factor of 1/100 for the ratio code/phase makes the precision of the estimated clock driven by the more accurate but ambiguous carrier phase, whereas the absolute accuracy is provided by the pseudorange. Several effects such as multipath, group delay or antennas center variations are averaged over the entire arc, but in practice these effects may not possess zero mean, and biases from arc to arc may be introduced.

5.3.3 Clock combination in IGS

In IGS, each analysis center computes a clock solution against a different time scale with a different software package and strategy. In order to generate the combined final clock product (stored as 'igswwww.clk'), a combination is performed by a weighted average of each estimate at each epoch. The quality of the combination is reported in clock summary files ('igswwww.cls'). The combination process requires several steps [86, 149]:

1. **Time Scale alignment**: In a first step, all analysis center clock products are aligned by a linear model $(a_0 + a_1 t)$ to the same time reference (e.g. satellite broadcast time or to reference center). For the alignment, all common satellite and station clocks can be used. The alignment process, adding the same correction to all clocks for one epoch, does not touch the internal quality (clock differences) at the epoch and does not influence the usage of the clocks (e.g. for PPP).

2. **Radial orbit correction**: As explained in Section 5.3, the clock estimates and the radial component of the orbit are strongly correlated. The epoch-wise differences between each

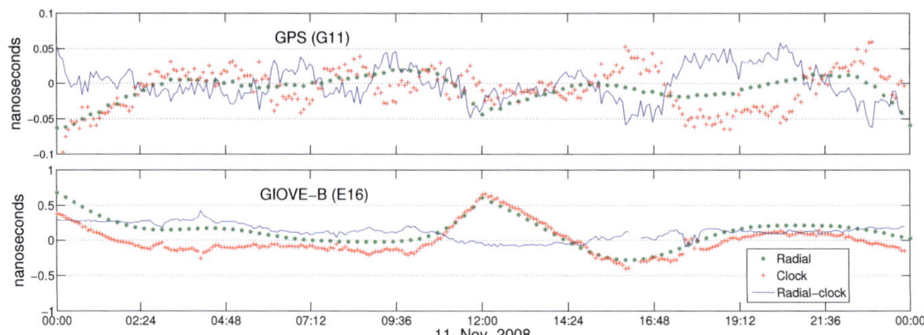

Fig. 5.8: GGSP. Clock and radial orbit difference between GFZ and ESOC.

analysis center with respect to the combined orbit ('igswwww.sp3') are computed, projected in the radial component and used to unify the clock inputs. This correction removes systematic biases introduced by the orbit and aligns final orbit and clock products. As a result 'unified' clock products can enter into the combination.

To understand this correction, it is useful to plot the radial orbit and clock differences between two processing centers for a Galileo and a GPS satellite. Figure 5.8 presents the difference between GFZ and ESOC estimations on 11 November 2008. For this particular day, the radial and clock difference indicated a clear correlation amounting to 1 ns differences for GIOVE-B. However, the blue line representing the clock agreement corrected for the radial difference shows a much more stable behaviour.

3. **Combination**: Combination is an iterative process, which combines all clocks at one epoch to get the best mean clock estimates for all stations and satellites. This process identifies outliers and jumps in the input products.

4. **Time scale creation**: Final combined clocks are aligned to an IGS time scale generated by the products themselves with a quality of about 1E-15. From this process, epoch-wise corrections are obtained, which are added to all clocks at the given epoch (no influence on PPP).

5.4 Ionosphere-free measurements

The main reason for transmitting more than one signal frequency is the dispersive behaviour of the ionosphere on the L-Band frequencies used for navigation. An ionosphere-free observable can be derived from the combination of two separated frequency measurements (code or phase) which eliminate the first order term of the ionosphere. Second and higher order terms, amounting to a few centimetres [17], are currently neglected. This combination becomes the basic input observation for the POD adjustment. The construction of the ionosphere-free linear combination needs to be reviewed to understand how different frequency dependent parameters are combined.

Let us start from the basic propagation equations for code and phase. The signal is generated on-board the satellite in the signal generator unit, broadcast by the antenna, propagated over the free space and Earth's atmosphere until it arrives at the receiver antenna where the signal is recovered and provided to the receiver, which then measures the propagation time delay against its replica of the signal. During its propagation path, several biases will delay the signal from the constant speed of light in vacuum.

In the next equations, the terms (DELAY_{rk}^s) should be read as associated to satellite (s) and/or receiver (r) for frequency (k). For example, $M_{GIEN\,L1}^{G01}(tor)$ shall be read as the multipath for frequency L1 for satellite G01 at the time of reception (tor) at the receiving station GIEN.

For code measurements the propagation equation can be written as:

$$
\begin{aligned}
P_{rk}^s(tor) = \; & \left| \bar{X}_r(tor) - \bar{X}^s(tot) \right| + \\
& + GDV_{rk}(tor) - GDV_k^s(tot) \\
& + CCO_{rk}(tor) - CCO_k^s(tot) \\
& + dt_r(tor) - dt^s(tot) \\
& + GD_{rk}(tor) - GD_k^s(tot) \\
& + 40.3 \frac{STEC_r^s(tor)}{f_k^2} \\
& + T_r^s(tor) + R_{tor}^s(tor) + S_r^s(tor) \\
& + DLLrk^s(tor) + M_{rk}^s(tor) + i_{rk}(tor)
\end{aligned}
\qquad [5.26]
$$

for carrier phase measurement :

$$
\begin{aligned}
\phi_{rk}^s(tor) = \; & \left| \bar{X}_r(tor) - \bar{X}^s(tot) \right| + \\
& + PCV_{rk}(tor) - PCV_k^s(tot) \\
& + PCO_{rk}(tor) - PCO_k^s(tot) \\
& + dt_r(tor) - dt^s(tot) \\
& + GD_{rk}(tor) - GD_k^s(tot) \\
& - 40.3 \frac{STEC_r^s(tor)}{f_k^2} + \varphi_{rk}^s(tor) \\
& + T_r^s(tor) + R_r^s(tor) + S_r^s(tor) \\
& + \lambda_k N_{rk}^s(t_0) + \lambda_k \left[\phi_{rk}(t_0) - \phi_k^s(t_0 - \tau_r^s(t_0)) \right] \\
& + PLLrk^s(tor) + m_{rk}^s(tor) + i_{rk}(tor)
\end{aligned}
\qquad [5.27]
$$

where :

$\bar{X}$ =[x,y,z], vector of satellite or receiver coordinates in the selected reference frame (e.g.ITRF).

tor , time of reception at correlation of the satellite signal by the receiver in system time.

tot , time of transmission of the signal in system time. In a first order, it is approximately equal to the time of reception minus the propagation time.

r , receiver subindex indicates that the term is receiver-dependent.

s , satellite subindex indicates that the term is satellite-dependent.

k , frequency subindex indicates that the term is frequency-dependent.

P , code measurement.

ϕ , carrier phase measurement.

GDV , direction dependent code or group delay variation associated to the transmitting or receiving antenna.

PCV , direction dependent phase center variation.

CCO , code center offset.

PCO , phase center offset.

dt , clock offset.

GD , absolute group delay from signal generation to transmitting antenna or from receiving antenna to signal correlation. This term can also be modulation-dependent within the same frequency (e.g. P1 and C/A code measurements).

$STEC$, slant total electron content at the station in the direction of the satellite. Possible scintillation effects are considered as part of the DLL and PLL noise.

f , frequency in Hz for k.

φ , phase wind up.

T , troposphere delay in the slant direction.

R , periodic relativistic correction and other minor relativistic terms.

S , Sagnac effect (described in section 3.7.1).

t_0 , time of the first carrier phase measurement at the receiver.

τ , signal propagation time between receiver and satellite.

λ_k , wave length for frequency k.

N , integer carrier phase ambiguity.

$\phi(t_0)$, initial fractional phase ambiguity at receiver or satellite.

DLL , code phase noise.

PLL , carrier phase noise.

M, m , multipath noise for code (M) and phase (m).

i , interference.

Most of these components are explained in the different sections of this thesis or are well known to geodetic users. Nonetheless, the components affecting the noise of the measurements (i.e. DLL, PLL and interference) are further explained in Section 6.2. In order to simplify the derivation, the terms in Equations 5.26 and 5.27 can be separated into the geometrical distance $\rho = |\bar{X}_r(tor) - \bar{X}^s(tot)|$, frequency independent terms A,

$$
\begin{aligned}
A_r^s = \ & +dt_r(tor) - dt^s(tot) \\
& +T_r^s(tor) + R_r^s(tor) + S_r^s(tor)
\end{aligned}
\qquad [5.28]
$$

and frequency k dependent terms B.

$$
\begin{aligned}
B_{rk}^s = \ & +PCV_{rk}(tor) - PCV_k^s(tot) \\
& +PCO_{rk}(tor) - PCO_k^s(tot) \\
& +GD_{rk}(tor) - GD_k^s(tot) \\
& -40.3\frac{STEC_r^s(tor)}{f_k^2} + \varphi_{rk}^s(tor) \\
& +\lambda_k N_{rk}^s(t_0) + \lambda_k \left[\phi_{rk}(tor_0) - \phi_k^s(tor_0 - \tau_r^s) \right] \\
& +PLLrk^s(tor) + m_{rk}^s(tor) + i_{rk}(tor)
\end{aligned}
\qquad [5.29]
$$

To facilitate the derivation, only one direction (receiver to satellite) and two generic frequencies are considered. As a consequence, the receiver r and satellite ^{s} related indices are dropped and only frequency sub-indices $k = 1$ and $k = 2$ used. Finally, the derivation of the ionosphere-free combination will be performed only for phase ϕ measurements in several steps from [5.30]; as a matter of fact, the same derivation applies to code measurements.

$$
\begin{aligned}
\phi_1 &= \rho - \frac{40.3STEC}{f_1^2} + A + B_1 \\
\phi_2 &= \rho - \frac{40.3STEC}{f_2^2} + A + B_2
\end{aligned}
\qquad [5.30]
$$

First, the slant total electron content value is eliminated by multiplying by the square of the frequency and subtracting the equations.

	f1	E5a	E5b	E5	L2	E6	L1
f2	Hz*f0	1150	1165	1180	1200	1250	1540
E5a	1150	-	39.08	19.92	12.26	6.51	2.26
E5b	1165	38.08	-	39.58	17.40	7.61	2.34
E5	1180	18.92	38.58	-	30.25	9.19	2.42
L2	1200	11.26	16.40	29.25	-	12.76	2.55
E6	1250	5.51	6.61	8.19	11.76	-	2.93
L1	1540	1.26	1.34	1.42	1.55	1.93	-

Tab. 5.1: ionosphere-free K factors for each possible ionosphere-free combination with GPS and Galileo frequencies

$$f_1^2 \phi_1 - f_2^2 \phi_2 = (f_1^2 - f_2^2)(\rho + A) + (f_1^2 B_1 - f_2^2 B_2) \qquad [5.31]$$

Second, in order to maintain the frequency independent terms ρ and A unscaled, Equation 5.31 is divided by $(f_1^2 - f_2^2)$:

$$\frac{f_1^2}{f_1^2 - f_2^2} \phi_1 - \frac{f_2^2}{f_1^2 - f_2^2} \phi_2 = \rho + A + (\frac{f_1^2}{f_1^2 - f_2^2} B_1 - \frac{f_2^2}{f_1^2 - f_2^2} B_2) \qquad [5.32]$$

Finally, the usual representation of the ionosphere-free combination is obtained by replacing the constant terms by $K_1 = \frac{f_1^2}{f_1^2 - f_2^2}$ and $K_2 = \frac{f_2^2}{f_1^2 - f_2^2}$:

$$\phi_3 = K_1 \phi_1 - K_2 \phi_2 = \rho + A + \underbrace{(K_1 B_1 - K_2 B_2)}_{B_3} + \ldots \qquad [5.33]$$

The total electron content dependency has been eliminated in the final equation. Non-frequency dependent terms A keep the same magnitude as in the single measurements while frequency dependent terms B are multiplied by the K factors. All frequency dependent biases get lumped into a joint term, resulting from the combination of both (e.g. B_3), amplified by a factor inversely proportional to the separation between the frequencies as described in Table 5.1. Noise in the GPS ionosphere-free combination (with L1-L2) gets amplified by 1.55 and 2.55, while the noise of a possible Galileo E5a-E5b combination would get amplified by 38.08 and 39.08 making it unusable in spite of the longer wave length.

In POD estimations, the different effects in the propagation in Equations [5.26] and [5.27] are modelled or included in the estimation in [5.22], except for the group delay which is lumped into the satellite and receivers clocks as observed in [5.34]. As a consequence, the estimated clock offset dt' should be called 'ionosphere-free clock' or 'signal clock' since it additionally contains the group delay bias from the signal generation up to the phase center.

$$\begin{aligned} dt'^s(tot) &= dt^s(tot) + K_1 GD_1^s(tot) &- K_2 GD_2^s(tot) \\ dt'_r(tor) &= dt_r(tor) + K_1 GD_{r1}(tor) &- K_2 GD_{r2}(tor) \end{aligned} \qquad [5.34]$$

This circumstance becomes particularly significant when different 'ionosphere-free clocks' are obtained based on different signals or modulations by introducing biases between them (as acknowledged by IGS for the processing of GPS clocks [IGSMAIL-2744]) and periodic divergences due to the possible variations in the group delay associated to its thermal dependency (as observed for first GPS L5 transmissions [111]).

5.5 Group delay estimation

For orbit and clock estimations, the ionosphere effect is eliminated by ionosphere-free combinations and hardware delays lumped into the clocks estimates. Since the ionosphere-free clock dt' contains the group delay, a closer analysis of the group delay is required in order to understand the clock estimates obtained by POD.

Nevertheless, other applications are focused on a separation of these two parameters. Single frequency navigation users require a correction of the group delay difference between the ionosphere -free clock dt'^s and the clock observed with a single frequency measurement $dt_k^s + GD_k^s$. This correction is included in the navigation message as the Broadcast Group Delay (BGD). Ionosphere correction is also required for single-frequency users, and TEC models are transmitted in the navigation message. Besides navigation uses, other applications, such as radio-telecommunications, are highly interested in the estimation and the monitoring of the ionosphere behaviour.

Group delays are obtained as a by-product of ionosphere estimations performed by using a different methodology than explained in Section 5.3. The basic principle is explained in the following. Ionosphere and group delays are computed from the geometry-free combination. The geometry is eliminated by differencing equations 5.30 for code observations:

$$
\begin{aligned}
P_{r2}^s(t) - P_{r1}^s(t) &= 40.3\frac{STEC}{f_2^2} - 40.3\frac{STEC}{f_1^2} + B_2 - B_1 \\
&= (\gamma I_{r1}^s - I_{r1}^s) + B_2 - B_1 \\
&= (\gamma - 1)I_{r1}^s + B_2 - B_1
\end{aligned}
\qquad [5.35]
$$

where :

$I_{r1}^s = 40.3\frac{STEC_r^s}{f_k^2}$ is the ionospheric delay at frequency f_1. Note the relationship $I_{r2}^s = \gamma I_{r1}^s$

$\gamma = (\frac{f_1}{f_2})^2$ is the relationship between the squares of both carrier frequencies.

This equation can be divided by $(\gamma - 1)$ in order to compute the unscaled ionospheric delay values.

$$
\frac{1}{(\gamma-1)}(P_{r2}^s(t) - P_{r1}^s(t)) = I_{r1}^s + \frac{1}{(\gamma-1)}(B_2 - B_1)
\qquad [5.36]
$$

The geometry-free observation equation removes all effects on the phases and pseudoranges that are common to both frequencies (such as distance from receiver to satellite, clock offsets, tropospheric delay, etc.), but frequency-dependent effects, like multipath and the differential instrumental biases in the satellite and in the receiver, are still present in the ionospheric terms. The main frequency dependent terms in B with no zero mean included in the equation are the group delays. The final equation can be rewritten as :

$$\frac{1}{(\gamma-1)}(P_{r2}^s(t) - P_{r1}^s(t)) = I_{r1}^s + \frac{1}{(\gamma-1)}(GD_{r2} - GD_{r1}) + \frac{1}{(\gamma-1)}(GD_2^s - GD_1^s) \qquad [5.37]$$

The ionosphere and group delay estimation are performed from Equation 5.37. In order to separate the contribution of the ionosphere from the inter-frequency group delay biases, the ionospheric term is mapped onto a vertical grid and the term $\frac{1}{(\gamma-1)}(GD_2 - GD_1)$ estimated as single values. As it is not possible to unambiguously determine all satellite and receiver biases, one of them is selected as a reference and fixed to zero ($GD_{\text{ref}} = 0$) or a separately calibrated value, or a zero mean condition is used e.g. $\sum_{s=1}^{n}(GD_2^s - GD_1^s) = 0$. Further estimated satellite and station group delay biases are relative to this condition. For each day, the mapped ionospheric terms from all available stations are combined in a Kalman filter or least squares process, where the coefficients of the polynomial for each station are considered as a random walk stochastic process and the group delay biases are considered to be constant for the entire estimation period. Averaged multipath errors may be still present in the process. To reduce the effects of multipath, high elevation cut-off data are used.

Now, the satellite ionosphere-free clock in Equation 5.34 can be expressed in terms of γ by substitution of the terms K, in order to get the same nomenclature for the satellite ionosphere-free clock dt'^s and the single frequency clock dt_1' for f_1 required by a single frequency navigation user:

$$\begin{aligned} dt'^s &= dt^s + \frac{\gamma}{\gamma-1} \, GD_1^s - \frac{1}{\gamma-1}GD_2^s(tot) \\ dt_1'^s &= dt^s + \qquad GD_1^s \end{aligned} \qquad [5.38]$$

The relation between them can be established by the subtraction of both equations:

$$\begin{aligned} dt_1'^s - dt'^s &= GD_1^s - \frac{\gamma}{\gamma-1}GD_1^s + \frac{1}{\gamma-1}GD_2^s(tot) \\ &= \frac{\gamma-1-\gamma}{\gamma-1}GD_1^s + \frac{1}{\gamma-1}GD_2^s \\ &= \frac{-1}{\gamma-1}GD_1^s + \frac{1}{\gamma-1}GD_2^s \end{aligned} \qquad [5.39]$$

which can be reordered into the final relation which is identical to the estimated terms contained in equation [5.37]:

$$dt_1'^s - dt'^s = \frac{1}{\gamma - 1}(GD_2^s - GD_1^s) \qquad [5.40]$$

This estimated value is called the broadcast satellite group delay (BGD), as it is transmitted to the user through the navigation message. A single-frequency user receiver processing pseudo-ranges from the frequency f_1 or f_2 shall apply the following additional correction to the ionosphere-free clock correction dt'^s based on f_1 and f_2 combination:

$$\begin{aligned} dt_1'^s(t) &= dt'^s(t) + \quad\;\; BGD^s \\ dt_2'^s(t) &= dt'^s(t) + \;\; \gamma \;\; BGD^s \end{aligned} \qquad [5.41]$$

This concept is directly applied to the Galileo message. The same parameter (called T_{GD}) is also used in the legacy message on L1 in GPS [115], whereas new terms have been introduced for the new signals on L2 and L5. The relation between legacy and the new parameters are derived from the same basic principle as explained in [162].

5.6 Conclusions

In this chapter the methodology for the clock parameter estimation has been presented.

Section 5.3 briefly reviewed the basic principles of the Geodetic time transfer between Satellite (dt^s) and Receiver time (dt_r) to system time. Several important conclusions have been extracted in order to understand the revision of the clock estimation accuracy and precision performed in next chapter. The mathematical model with a kinematic approach for clock and a dynamic one for orbit makes both estimations strongly correlated especially considering the radial component; and both constituents need to be understood and analyzed together in order to assess the accuracy of the clock estimations. Additional correlations can exist alongside other estimated parameters (as satellite SRP or station ISB). The weighting scheme has been identified to make the time transfer *precision* be based on the 100 times more precise, but at the same time, ambiguous carrier phase; while the *accuracy* is provided by the unambiguous code observations. Finally, absolute estimation can present small biases from arc to arc, known as boundary clock jumps - if constant estimated parameters are not stable (e.g. ISB), or do not average to zero during the pass (e.g.multipath or group delay variations).

It has been highlighted how the vector of unknowns is dominated by the clock estimates which represent around 80-90 % of the overall unknowns in the estimation. A kinematic model is used for the clock, whereas a dynamic model is used for the orbit, and other parameters are estimated as constants or slowly varying quantities. The kinematic model for all clocks makes these quantities become the predominant unknowns. Frequency standards are constantly improving, as demonstrated in Section 4.2, in case of improvement of the satellite and receiver frequency sources, the possibility to use a dynamic model for the clock (based on 2-3 parameters) would drastically reduce the amount of unknowns and the correlation with other estimated parameters. PHM on board Galileo satellites and H-masers at ground stations could already provide a validation of this hypothesis. Nevertheless, the group delays currently included in the 'apparent' clock have to be carefully taken into account.

Section 5.4 has analyzed in detail the ionosphere-free basic modelling observation used in POD. It has now been demonstrated how current POD techniques, based on the linear combination of two frequencies to create the ionosphere-free measurement, lump the satellite hardware delay biases and instabilities into the estimated clock that as a consequence shall be called 'apparent' or 'ionosphere free' clock.

This term will be used hereafter to distinguish between the AFS noise ('physical clock'), signal noise at the output of the satellite antenna for each frequency ('signal clock') and the final clock noise derived from POD ('apparent clock'). It has to be remarked that the term 'signal clock' is also used by other authors to refer to the 'apparent clock' [120]. However, hereafter the concept 'signal clock' will be considered more appropriate as the different signals can have different behaviour.

In Section 5.5 has been constated that the separation between ionosphere and group delays is not an easy task, and requires zero mean assumptions or absolute calibrations of stations and satellites.

6 Performance of geodetic time transfer

6.1 Introduction

Once the satellite clock phase offset with respect to the system time (dt^s) is estimated by POD, it is used to perform time transfer to the users, to characterize the 'signal clock' performance and review system design parameters. However, before using the estimated clock phase it has to be clarified what is the quality, precision and accuracy of this estimation.

POD accuracy and precision is a continuously revised topic in the IGS community. The accuracy of IGS products is constantly analyzed by the analysis center coordinator, published on-line (under http://acc.igs.org/), reported by mail (e.g.IGSMAIL-6053) and reviewed at IGS-Workshops (e.g. Newcastle 2010). The IGS official accuracy on the product website description is 0.075 ns (rms) for clock and 5 cm (1D,rms) in the orbit component for final products. The precision for both products can be considered better with a value of 0.02 ns (1σ) [63]. However, for new satellite systems as Galileo the IGS does not provide any solution and it is necessary to review the accuracy and precision of the clock estimates from scratch.

This chapter analyzes the precision and accuracy of GNSS time transfer by reviewing the quality of the clock estimations performed in GIOVE mission for GPS and Galileo satellites. Methodology and results for GPS will be cross-checked with IGS. Precision and accuracy of geodetic time transfer will be derived in a step-wise approach. First, the internal precision is analyzed from fitting residuals and repeatability of results. Second, the external precision is checked against other software packages with different algorithms, different data networks and known reference measurements. Finally, the absolute accuracy will be provided by comparison against independent techniques such as Satellite Laser Ranging for the orbit and Two-Way Satellite Time and Frequency Transfer (TWSTFT) for the signal clocks for stations located at timing laboratories.

6.2 Theoretical limit: code and carrier phase quality

In the receiver the code measurement is performed by a Delay-Lock-Loop (DLL) and the phase by a Phase-Lock-Loop (PLL). The accuracy of both tracking loops can be computed following [22] for the code (DLL) tracking :

$$\sigma_{DLL} = \lambda_P \sqrt{\frac{B_{DLL}d \cdot K_{BOC}}{2C/N0}\left(1 + \frac{1}{T_{DLL}C/N0}\right)} \qquad [6.1]$$

and for the phase (PLL) in meters :

$$\sigma_{\phi_k} = \frac{\lambda_\phi}{2\pi} \sqrt{\frac{B_{PLL}}{C/N0}\left(1 + \frac{1}{2T_{PLL}C/N0}\right)} \qquad [6.2]$$

where :

B_{DLL} is the code loop bandwidth (Hz)

B_{PLL} is the phase loop bandwidth (Hz)

λ_P is the code chip length (m)

λ_ϕ is the carrier phase wave length (m)

T is the loop predetection time for DLL or PLL (s)

d is the DLL correlator chip spacing

K_{BOC} is a factor for BOC signal performance

$C/N0$ is the signal to noise power ratio expressed as a ratio $10^{CN/10}$ (dB-Hz)

From Equations 6.1 and 6.2 follows that the measurement accuracy depends on the received power $(C/N0)$ and the exact configuration of the receiver parameters (B, T, d). Once the receiver configurable parameters are known, it is possible to predict the carrier and code phase noise expected for a determined C/N0 value. For the static GETR receiver used in GIOVE mission, the configurable values are:

- $B_{DLL} = 1.0$ Hz

- $B_{PLL} = 5.0$ Hz

- $T_{DLL} = 0.1$ s

- $T_{PLL} = 0.01$ s

- $d_{E5} = 0.358, d_{E6} = 0.179, d_{E1A} = 0.0895, d_{E1B} = 0.0358$

Figure 6.1 presents the expected noise in seconds versus C/N0 for the above configuration for all GPS and Galileo signals. As observed, the signal to noise ratio is the key factor to obtain a more accurate measurement.

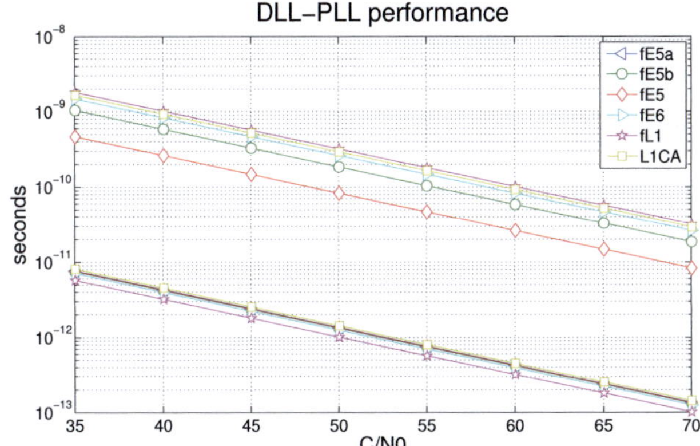

Fig. 6.1: Theoretical DLL and PLL noise for GETR configuration values

The validity of Equations 6.1 and 6.2 can be analysed with the GIOVE mission network. In order to cover different periods of GIOVE-A double frequency transmission, two periods of 21 days are selected during March (E1-E5) and (E1-E6) September 2007. From the latest period just the 5 initial days are taken for GPS due to the higher number of satellites and data available. All 13 GESS stations are in the analysis and a mean value is obtained for GIOVE-A and the complete GPS constellation. The GESS receiver has a lower configuration limit of 30 dB from which the signal start to be tracked independently of the elevation. The expected noise figures using Equations 6.1 and 6.2 with the reported C/N0 values are compared against the measured ones.

In Galileo, besides the data channel, several signals include a pilot component without data modulation in order to facilitate the acquisition with low C/N0. The subtraction of pilot and data measurements (e.g. $\phi_{E5aI}(t) - \phi E5aQ(t)$) eliminates all effects except the tracking noise ,which is incremented by $\sqrt{2}$, and renders possible a clear analysis of the tracking noise for the carrier phase. In Figure 6.2 a good agreement is observed between measured C/N0 values and the real receiver performance using pilot minus data combinations. The dispersion between the different stations was only significant for code measurements for the lowest C/N0 values (30-35dB) with maximum differences of 1 dm with respect to the mean.

Figure 6.2(c) presents the C/N0 versus elevation for Galileo and GPS signals obtained as the mean value for the reference period. Different power levels are observed for each signal. The agreement between the stations was ± 1 dB with the exception of La Plata (Argentina) were the measured values were -5dB with respect to the average for all frequency bands. This was believed to be linked to the high multipath observed for this station.

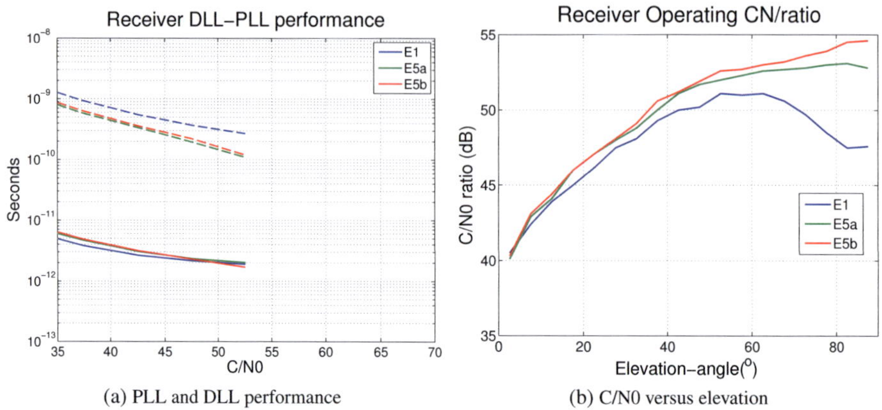

(a) PLL and DLL performance   (b) C/N0 versus elevation

Fig. 6.2: Measured receiver code (DLL), carrier (PLL) and signal-to-noise ratio (C/N0) in GIOVE mission network

The computation of the signal to noise ratio varies depending on the receiver manufacturer and receiver antenna gain. Effective C/N0 measurements at the receiver for a given signal depend on the effective isotropic power radiated by the navigation antenna, the propagation loss, the gain of the receiving antenna and the receiver architecture. This characterization is only valid for a given type of station, as it is also observed in [68] by comparing different multi-GNSS receivers.

Possible interferences should not be ignored as their importance is rising in parallel to the increased occupation of the radio frequency spectrum by other satellites or terrestrial services. The weak spread spectrum signal received from the satellite is raised over the noise floor by autocorrelation with the receiver code replica. The final signal to noise ratio (C/N0) is measured by the receiver with respect to the noise floor. The Equations 6.2 and 6.1 depend on this value. Any interference (last term in Equation 5.26 and 5.27) will affect the quality of the received power (C/N0) and, in consequence, the noise of code and carrier phase measurement.

The interference may be divided in satellite and ground sources. For the satellite source, many publications addressing the radio frequency compatibility between Radio Navigation Satellite Signal have been generated in recent years to account for the increasing number of GNSS and SBAS satellites. For example [166] represents a very useful reference and the methodology is agreed at ITU level and described in RTCA 1831 [79]. For the second type, the deployment of the terrestrial LightSquared system in the United States brought significant attention to the ground sources leading to the non-final authorization to operate on the border of L1 band.

Indeed, the current band allocation in Figure 6.3(a) deserves some attention when looking to ground interferences: L1 band centered on 1.575 GHz is allocated only to Radio Navigation

Satellite Systems (RNSS); L5/E5 spectrum band centered on 1.207 GHz is however shared with air traffic control systems (DME and TACAN) used for Aviation and its influence on airplanes is not negligible; and L2 (1.2276 GHz) and E6 (1.278 GHz) bands also shared with radio-amateur transmission. The authorised radio-amateur transmission can generate important interferences, as it was the case for the GIOVE station located in Turin (IT) in Figure 6.3(b). Additionally, unintended out-of-band transmissions from operating systems in other bands can cause severe degradation of the measurements, as it was also the case for he same station due to a local out-of band spurious TV emision until it was localized and fixed.

In summary, once the C/N0 for a station is characterized, it is possible to compute the expected C/N0 for a satellite at a given elevation and predict the expected noise for a signal. Following this approach the expected accuracy of the time transfer measurements without multipath contribution may be predicted by knowing the satellite receiver geometry at a given instant. Time transfer with 10-1 ps precision should be possible using phase and with 1.0-0.1 ns accuracy using code measurements. Nevertheless, a special attention has to be given to the environment to avoid multipath and interference sources.

6.3 Precision and accuracy of geodetic time transfer

6.3.1 Measurements residuals

A first indicator of the quality of POD results are the measurement residuals, i.e. the difference between real data and the measurements as modelled by the processing algorithms. The residuals are expected to be zero mean, randomly distributed and in agreement with the expected noise contributions, namely, thermal noise values for DLL and PLL (presented in section 3.6.1) plus multipath. Typical residual RMS values obtained with the GIOVE mission are 40 cm for code measurements, 1 cm for phase observations and 3 cm(one-way) for SLR [167].

The detailed analysis of the residuals for systematic effects can provide additional information on the adjustment quality. For example the standard deviation of the residuals can be related to elevation. In such cases, a systematic elevation-dependent pattern of the mean values is observed on code residuals for both GIOVE and GPS in Figure 6.4(a). This effect is not present in the phase measurement and is due to a code/phase incoherence (also called group delay) versus elevation associated to the station Spaced Engineering antenna type-1. This effect was considerably reduced with the installation of a modified version of the antenna. This type-2 antenna presents zero mean code residuals versus elevation as observed in Figure 6.4(b).

In addition, the standard deviation is also elevation dependent in both cases in correspondence with the effective received signal-to-noise ratio (depending on the transmitting antenna power, propagation losses and receiver antenna gain). If the receiver logs the signal-to-noise ratio this

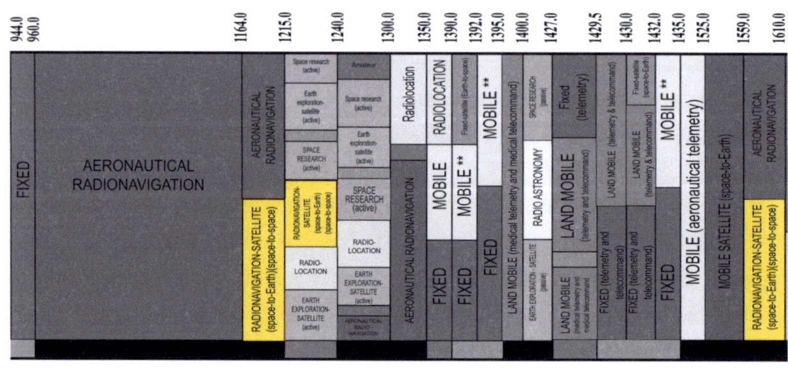

(a) United States frequency allocation in L-band. Source: Wikipedia

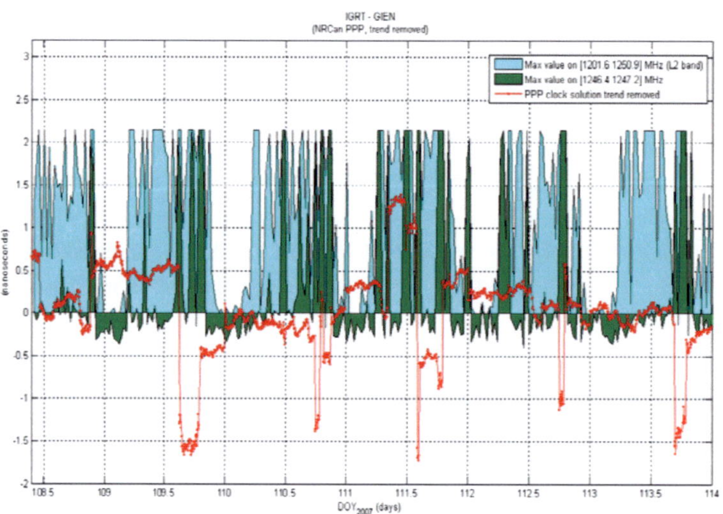

(b) Spectrum analyser versus clock estimations at GIEN station. Source: INRiM

Fig. 6.3: Example of local interference in L-band in clock estimations

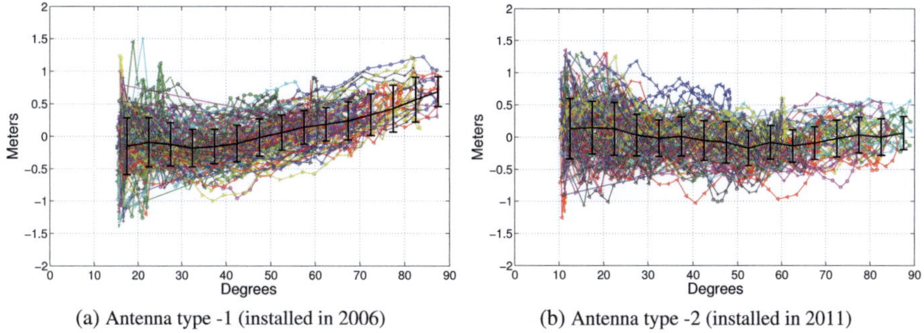

Fig. 6.4: Code residuals versus elevation for initial Type-1 and final Type-2 Space Engineering antennas installed at New Norcia, Australia

measurement can be used to compare with the expected values and extract further conclusions as in Figure 6.2 where a good agreement with the theory in subfigures (a) and (b) can be observed.

The analysis of the residuals is also performed by IGS analysis centers in the summary files (*.sum) of the processing. In these files the residuals are presented by station and satellite. Analysis centers using double differences (e.g. CODE) report the residuals at the stations by baseline. Typical values for GPS satellites extracted from esa*.sum are 1 m (rms) for code and 10 mm (rms) for phase, or 3 and 0.3 ns in units of time, in line with the typical weight factor of 100 applied for the observations in the POD adjustments.

6.3.2 Repeatability of results

The orbit and clock estimation is performed by 'arcs' which cover a period from 1 up to 10 days. In case that the same period is covered by different arcs, it is possible to check the internal accuracy by comparison of the different solutions. The consistency of the estimations in the overlapping interval is used as the main performance indicator.

In GIOVE mission five days arcs are used with one day overlapping at the boundaries, as graphically depicted in Figure 6.5. The orbit difference vector is projected into the so-called worst user location (WUL), i.e. the point on the Earth's surface where the projection of the orbital error is maximum, then the RMS over the 1-day overlap period is computed. The global RMS obtained from orbit restitutions is 14.3cm for GIOVE-A and 14.5cm for GIOVE-B. Due to the limited tracking network, the overlap results during eclipse periods are slightly worse, more precisely 22.4 cm for GIOVE-A, whereas for GIOVE-B, the orbit RMS is 17.8 cm during eclipse periods.

The observed differences between eclipse/non eclipse orbit predictability are believed to be caused by the difficulty of estimating precisely the solar radiation pressure (SRP) parameters due to the reduced GESS network, which prevents an adequate observability of the satellite dynamics. As the SRP model applied is purely empirical, no a-priori information about the SRP parameter values is known, so they can fluctuate arbitrarily [167].

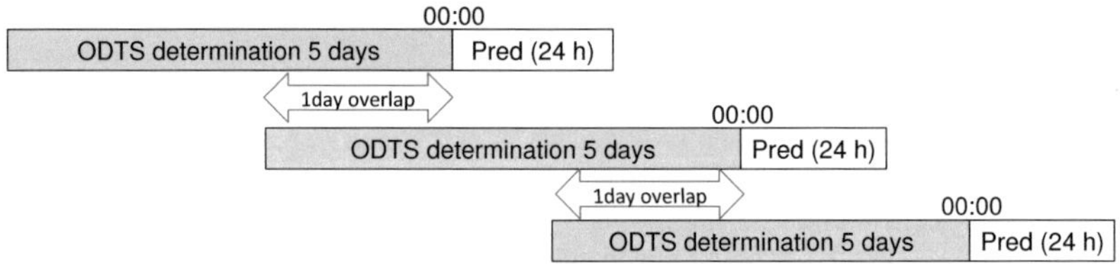

Fig. 6.5: Overlapping arcs

Regarding the clock overlap results, the overall precision for GIOVE-A is 0.56 ns (rms), whereas for GIOVE-B this value decreases to 0.51 ns. It is important to point out, that the RMS is similar for both satellites which is a normal result since clocks are estimated as snapshots; that means, no specific model is applied when estimating the clock parameters, as they are estimated on an epoch per epoch basis. For the same reason there is also no noticeable difference when using different types of on-board clocks (RAFS or PHM). The standard deviation is stable around 0.30 ns (1σ) for both clocks.

The clock overlap error is slightly larger during eclipse seasons for GIOVE-A (0.73 ns rms), while it is almost equal for GIOVE-B (0.52 ns rms). This can be explained by the lower precision in the orbit restitution during eclipses. Additionally, the attitude modelling may be an additional error source for clock restitution during eclipse. The attitude law becomes singular during the eclipse season requiring faster turns of the satellite around the Z axis (yaw) to keep its pointing attitude. The inaccurate modelling of the noon turns is mainly absorbed by the clock, as observed for GPS Block IIA satellites [84]. This error source is unlikely for GIOVE-B with minor effects on the clock estimations as observed later in Figure 7.11 but it could be a reason for the higher rms in GIOVE-A since no singular model was implemented in the POD processing software for this second satellite.

A similar strategy is performed in IGS for the orbit combination reported in the summary combinations files (*.sum) by some of the processing centers by reporting the orbit overlap with the previous day. For example GFZ reports an orbit repeatability of 8-9 cm (1σ,1D). Satellite on eclipse conditions are also distinguished, as poorer results are obtained for eclipsing satellites.

Arc boundary jumps

In case no overlapping periods exist, as in the case of final IGS clocks products, it is possible to check the differences between the clocks at the border days in a similar way as performed for the orbit in [63]. To facilitate the comparison H-maser stations are normally used. Some of the stations in the network use H-maser clocks with a nominal stability of $10^{-13}\tau^{-1/2} + 2 \times 10^{-16}$ [155]. The difference between nominal 5 min interval samples for this kind of clocks is expected to be in the picoseconds range. Clock estimation of stations with H-masers is applied by IGS to analyze the accuracy of clock products and stations performance. The boundary jumps between data batches provided on-line under https://goby.nrl.navy.mil/IGStime/daybdy/ offer a good estimation of the quality of the clocks.

The RMS boundary jumps in IGS show a large dispersion ranging from 0.1 to 1 ns [140]. Several causes are mentioned in [140] and [38], linked to changes in the hardware delays due to temperature variations, damages or equipment changes at the station. As station, the chain of all elements which contribute to achieve a measurement is understood: antenna, cables, possible splitters, receiver, external frequency source and the hardware/software in charge of controlling the station. In POD solutions phase observations are normally weighted 100 times higher than the code, nevertheless code data are necessary to recover absolute time transfer. In practice, for each estimation arc, the absolute clock offset is determined by the mean code value of the observations and the carrier measurements provide the relative clock. Consequently, any alterations in the absolute code values (equipment change) or the quality of the code observations will affect the boundary results.

Particularly interesting is the case of the Canadian stations depicted in Figure 6.6. Even if one of the lowest value for boundary jumps is observed for NRC1, all stations present seasonal variations from 0.1 ns to 2.0 ns. These variations seem to be coincident with the snow fall in the region typically covering the period from October to April. For example, at YELL, the increase in the RMS values at the end of 2010 starts associated with the first snow fall reported on 14th October. Other stations located in similar latitudes, as NYAL in Norway, do not present such seasonal variations. One reason could be linked to the accumulation of snow on the antenna as the three Canadian stations have choke ring antennas without dome protection whereas the

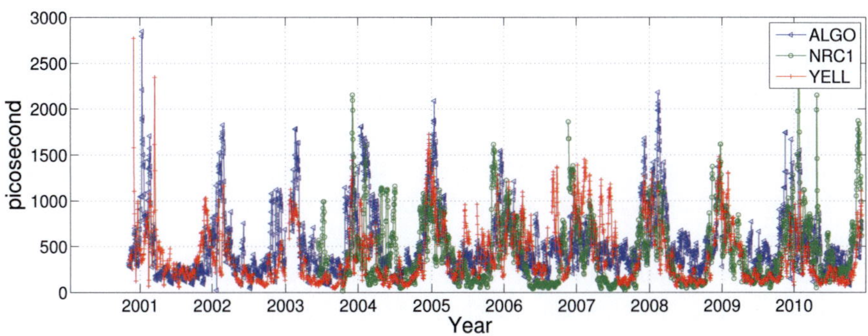

Fig. 6.6: RMS of daily boundary jumps for a subset of Canadian stations with H-masers

Norwegian antenna includes a dome. Physical explanations could be related to changes in the electrical properties of the antenna or the additional delay caused by the accumulation of snow.

6.3.3 Precision against reference H-Masers standards

Estimations of station clocks with H-masers can also be used to verify the clock stability by comparing the obtained Allan deviation against the specifications or against a better standard available at the laboratory. It should be emphasized that estimating the on-board clock or signal clock observed from the ground is more difficult than measuring the same clock located on the ground. The apparent clock is sensitive to the POD method used and the error sources that limit the estimates.

Figure 6.7 presents the Allan deviation for a subset of IGS stations for a 6 months period, the difference between IGS-GPS time scales, the specifications of the space PHM and two principal commercial H-masers, Symmetricom [161] and T4Time [155]. The Allan deviation for the stations is higher than expected from the pure atomic frequency standard which indicates first that the real frequency source stability is not preserved by the station electronics and second that the POD estimated measurement noise is higher than the H-masers and also the PHM. The results for this long period are slightly worse than shown in Figure 6.9 in section 6.3.4 which has been obtained for a limited short period and represents the best observed result.

In conclusion, time transfer precision by POD during short periods can provide noise levels of 1E-12 $\tau^-1/2$. The accuracy is however worse due to day-boundary discontinuities in the geodetic time transfer solution. This noise level should be considered as the geodetic time transfer system noise and therefore as the lower limit below which no on-board clock estimation is possible.

6.3.4 Clock validation by TWSTFT

Several of the H-maser clocks estimated by POD in IGS and two of the GIOVE mission are located in timing laboratories. Such clocks cooperate in the creation of UTC by BIPM and are already compared by different GNSS techniques, as GPS time transfer (POD, PPP or common view) or totally independently, such as Two-Way Satellite Time and Frequency Transfer (TW-STFT) explained in Section 5.2.2. Such clocks are also compared to other in house clocks with similar or better quality. For example, Figure 6.8 shows how the H-Maser connected to the GESS station in Turin (I) is locally compared to the UTC(I) realization performed at the timing laboratory.

Clock phases are estimated by POD with respect to a reference point which is the antenna phase center, while typically in time metrology the measurement reference point for a clock or time scale is physically located inside the laboratory at a certain point in the time scale genera-tion chain. The difference between a signal clock restituted at the center of phase of an antenna

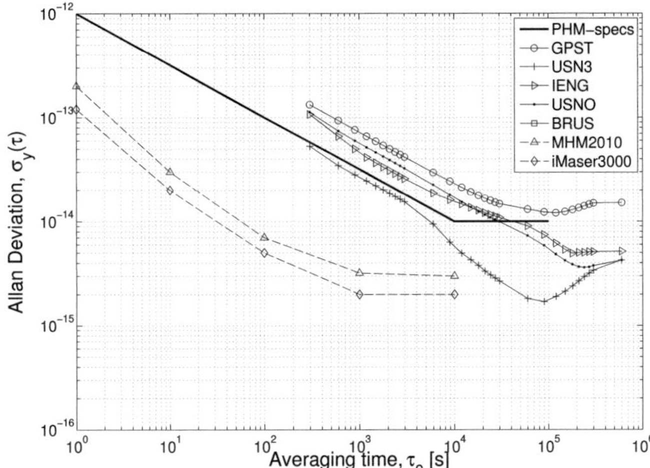

Fig. 6.7: Allan deviation for a subset of IGS stations with H-masers. Specifications of the two main ground commercial masers from Symmetricom [161] and T4Time [155] have also been included.

and a reference point inside the lab is given by the delays and instabilities of the receivers, antennas, cables, residuals in the estimations, among others, and the calibration of these delays asks for a careful procedure. Absolute time transfer requires the calibration of these values which are usually lumped into the estimated ionosphere-free clock.

To compare the precision of time transfer by GPS(PPP) and TWSTFT and to evaluate the noise added by the measurement system several tests have been performed. The timing laboratories INRIM(I) and USNO(US) have been selected as they host GNSS stations as part of IGS and GIOVE mission networks. The IGS stations IENG and USNO are connected to H-masers which are steered to UTC(I) and UTC(USNO) respectively in order to create a physical realization of UTC(k).Both institutions are equipped with TWSTFT links.

Additionally, NRCan PPP software is used at INRIM time laboratory in charge of UTC(I). Further details on the PPP algorithms, models and specifications implemented in NRCan can be found in [85]. NRCan's implementation of the PPP method was originally developed as a geodetic tool to provide single-station positioning capability within geodetic reference frames. NRCan PPP clock solutions achievable by means of PPP are consistent at the 2 nanosecond level with TWSTFT measurements [122]. PPP results showed a two-fold improvement in stability over two traditional GPS time synchronization methods (single and dual-frequency common view GPS), providing a frequency stability (in terms of Allan deviation) of 1E-14 over an averaging period of one day.

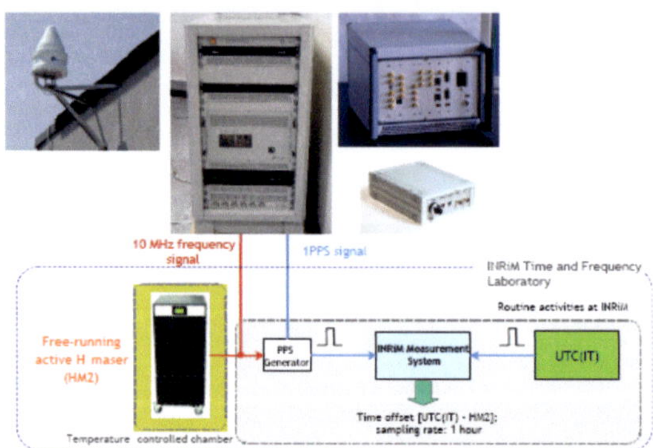

Fig. 6.8: UTC(I), INRIM (Italy), Time Laboratory set-up for GESS station [32].

Discontinuities in the order of 1 ns are observed between batch solutions. In the standard PPP solution the pseudorange is weighted around 100 time less than the more accurate carrier phase. Nevertheless, only the code allows for an absolute time transfer, the discontinuities being considered to be linked to the pseudorange noise averaging [38].

Figure 6.9 shows that the obtained POD-IGS results are comparable to the results provided by PPP (as expected) and TWSTFT. The stability of the active H-maser at IENG compared to the local cesium primary standard fountain is at the level of a few units in 1E-13 at $\tau = 1$ sec, reaching the level of 1E-15 at $\tau \approx 10^5$ sec.

The grey lines in Figure 6.9 represent the typical carrier phase noise slope ($1\text{E-}11\,\tau^{-1}$) for the ionofree combination and the expected PHM performance ($1\text{E-}12\,\tau^{-1/2}$). They are introduced as reference for a better comparison with Figure 6.7 with which it presents a good agreement. The carrier phase noise is well below the results obtained with POD or PPP what indicates that other limitations are present in the time transfer technique rather than the noise of the measurements (PLL), hampering the observation of the true H-maser or even the PHM in orbit.

Typically the remote comparison of two H-masers of this kind with the state of the art time transfer methods does not allow observing this H-maser stability because the noise of the measuring system is dominating. The comparison of station clocks with H-masers can then allow an estimation of the noise injected by the POD system when estimating better ground clocks. It should be emphasized that the ground clock estimated by GNSS techniques includes also the contribution of the station noise (DLL,PLL) and group delay thermal sensitivity. Indeed the increase observed in the Allan deviation around $\tau = 4000$ sec was not due to the H-maser but due to the thermal variation of the IENG station which was not visible in the direct comparison to the cesium fountain. The effect disappeared later after the receiver and location were thermally stabilized.

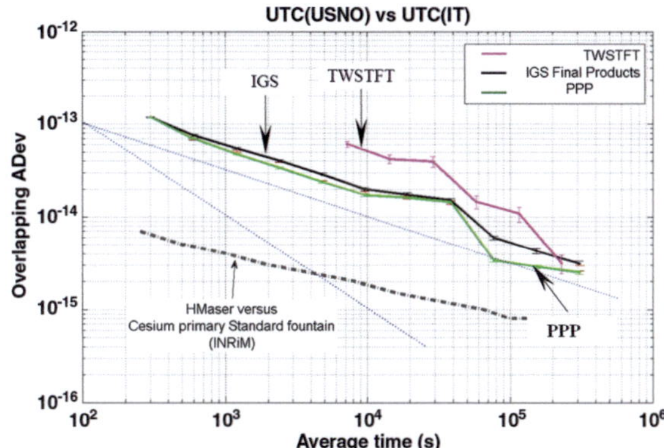

Fig. 6.9: Allan deviations of two H-masers compared using different techniques, namely: the two-way satellite time transfer, IGS final products and Precise Point Positioning [170].

6.3.5 Reproductivity by independent results

Once the internal consistency has been demonstrated by the analysis of residuals, overlappings or the stability of stations with H-Masers, the next question is whether an independent software or ground network achieves the same results. Since GPS and GIOVE satellites are estimated together, the first step is to check the accuracy of GPS estimations performed by the ODTS against final IGS products normally used as benchmark in any GPS estimation. Results are presented in Figure 6.10. For a better comparison, the differences are also expressed in time domain for the radial component. Values amount to 0.1 ns (1σ and rms) for the radial component and to 0.6 ns (rms) and 0.25 ns (1σ) for the clock solutions. Both outcomes are consistent with the overlapping results.

Unfortunately no results have been generated for GIOVE by the IGS. The extrapolation of GPS accuracy results to GIOVE is not straight forward as some difference may exist in the modelling of both satellite families and the number of flying satellites is significantly different. However, a straight forward comparison may be performed against the GIOVE estimations performed by the GGSP with GESS network or by DLR with CONGO network. Both GGSP and CONGO have been presented in Section 2.5.

The GGSP orbit quality is extracted from the *.sum files with 0.150 m (rms,3D) typical values for the combination. Clock combination results presented by the GGSP consortium (IGS Workshop, Miami, June 2008) provided a precision of 0.02-0.05 ns (1σ) against the final clock solution. Daily overlaps however indicated a degraded accuracy of 0.5 ns (rms) supposed to be due to the ISB instability. Additionally, a further independent assessment is provided by a fully separated network as CONGO with typically 1-9 cm (rms,3D) values for orbit estimations depending on the arc length and SRP model [157].

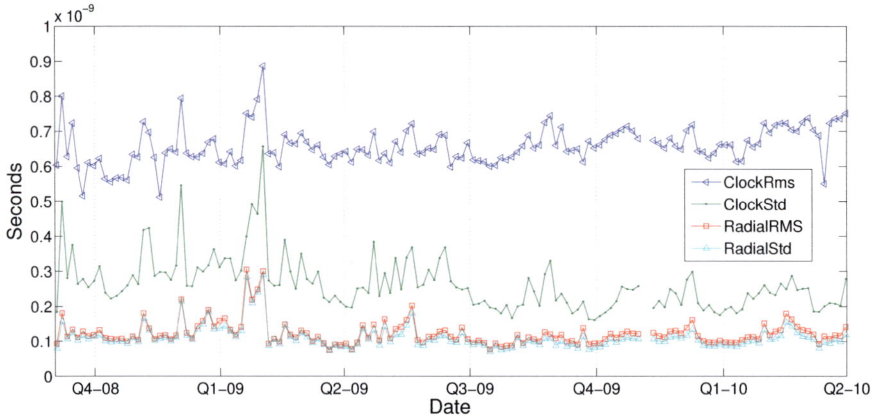

Fig. 6.10: ODTS clock and radial orbit component accuracy versus IGS

Before comparing independent clock solutions special attention has to be given to the antenna phase center offset used in the processing. The apparent clock may absorb any inaccuracy in the modelled effects as part of the processing. This is particular true for the phase center offset of the satellite antenna. As highlighted in Figure 4.8, unmodelled offsets between the center of mass of the satellite (reference for orbit estimation) and the center of phase of the navigation antenna (from where signals are radiated) are easily absorbed by the clock once projected into the radial component. Since the main component of the phase offset is the Z axis (center of mass to Earth direction) the use of different offsets will be easily absorbed by the clock.

This is clearly visible by comparison of different clock solutions in Figure 6.11 versus IGS. Constant offsets are observed between solutions, some particular satellites being out of the group. In the subfigure (a), the ODTS offset for one particular satellite (G05) was due to a missing update of the satellite antenna offset file (igs05.atx) after a PRN was reassigned. In the subfigure (b), the GFZ offset for a different satellite (G07) is not clear.

6.3.6 Precision dependency on number of sensor stations

It remains to understand the difference between the 0.25 ns (1sigma) and 0.05 ns (1sigma) achieved for GPS satellites by GIOVE mission and IGS respectively. The main hypothesis is the low number of sensor stations available in GIOVE mission. In order to analyse the relation with respect to the number of stations, NAPEOS software has been run with an incremental step to assess the influence of the number of stations on GPS satellite products. For each subset N of stations a test has been performed with the following configuration: N number of IGS stations globally and uniformly distributed; reference period of 5 days from 2010/02/19-24;

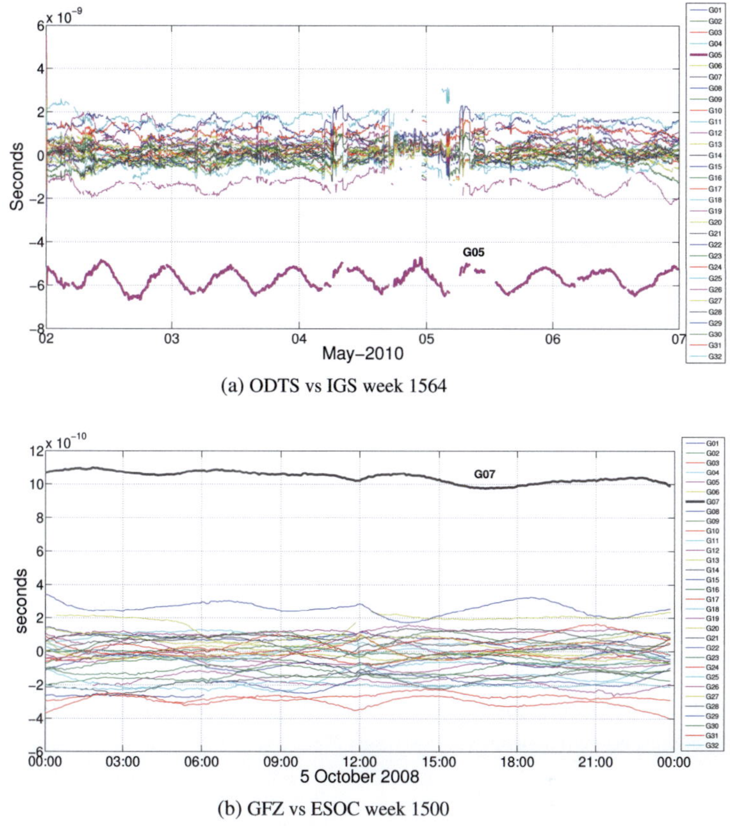

(a) ODTS vs IGS week 1564

(b) GFZ vs ESOC week 1500

Fig. 6.11: Clock offsets between processing centers

fitting interval of 1 day; separation of 1 hour between batches leading to a total number of 120 batches; results computed as 1σ, 1 axis (3D/$\sqrt{3}$) standard deviation of the differences of all GPS satellites orbits wrt to IGS final positions.

The final result is presented in Figure 6.12. The precision and dispersion improves with a ratio $1/N$ until N=28 stations. It can be concluded that the 10 times lower precision in GIOVE mission with respect to IGS and the higher noise observed for the POD apparent clock is in line with the limited network of 13 stations.

6.3.7 Orbit validation by SLR

POD time transfer absolute accuracy can be validated against the independent TWSTFT technique as explained in section 5.2. On the other hand, POD orbit absolute accuracy can be veri-

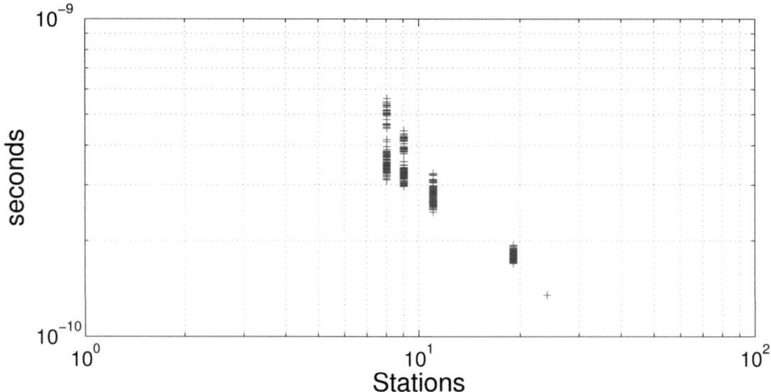

Fig. 6.12: Orbit accuracy versus number of stations

fied against independent Satellite Laser Ranging (SLR) measurements. SLR measurements are two-way ranging measurements between ranging stations and the satellite retroreflectors. The International Laser Ranging Service (ILRS) provides global satellite and lunar laser ranging data and their related products to support geodetic and geophysical research activities [125].

SLR observation residuals for GIOVE satellites as estimated by the ODTS are typically 5 cm (1σ, one-way) and 15 to 25 cm (rms). Details by satellite are provided in Table 6.1. The values are in agreement with the values observed for the clocks where differences of 0.3 (1σ) and 0.6 ns RMS were observed against IGS products. An offset of unknown source is observed which is also confirmed by independent estimations [164, 157].

	rms	mean
GIOVE-A	0.233	-0.032
GIOVE-B	0.131	0.082

Tab. 6.1: SLR residuals (cm)

6.4 Precision and accuracy of group delays estimation

6.4.1 Inter-frequency biases

As highlighted in Section 4.4, the presence of hardware delays is well-known and calibrated by the satellite manufacturer on ground. In order to compute the broadcast group delay (BGD) transmitted by the satellite for the single frequency user (explained in the previous section 5.5) calibrated values were initially used by GPS. However, discrepancies were observed between the calibrated values and the ones obtained by network adjustments. The values computed using network adjustment by the Jet Propulsory Laboratory are finally transmitted by the satellite for single frequency users. The network adjustments require some kind of zero mean assumption

	GIOVE-A		GIOVE-B	
	E5	E6	E5	E6
E1-	888.45	-5.81	881.18	3.96
E5-	0	-894.27	0	-877.22

Tab. 6.2: Inter-Frequency Biases for GIOVE-A and -B from calibration for nominal chain and pilot signals. Values in nanoseconds

(e.g. $\sum GD_{rk} = 0$) which is not realistic especially for small networks and the estimation of the ionosphere with additional assumptions.

Several advances allow the revision of this approach. If one receiver can be absolutely calibrated it could be used as reference, triple carrier transmissions in new GNSS satellites and the use of high gain antennas allow for additional frequency combinations with some new proposed methodologies. GIOVE satellites provide an interesting opportunity to review the different group delay estimation methodologies.

Absolute Satellite Calibrations

Absolute calibrations of satellite group delays can be performed as the sum of the single calibration of the different units and cables. Delays from navigation signal generation output to satellite antenna input can also be calibrated by introducing a test signal. Calibration can be performed at different temperatures and the thermal dependency variation can be extracted. This methodology is the only procedure to compute absolute GD_k^s. In order to compare with the other methodologies the group delay difference between frequencies, the so called Interfrequency Biases (IFB) or differential code delay (DCB), will be used:

$$IFB_{2-1}^s = GD_2^s - GD_1^s \qquad [6.3]$$

Values reported in Table 6.2 provide the IFBs for the different frequencies on the nominal chain for GIOVE-A and -B [135]. The similar difference between group delays for both satellites is justified by the delay in the frequency generation and up-conversion unit (FGUU) which is the only common unit in the navigation chain after the signal is generated in the navigation unit, as graphically depicted in Figure 4.7.

Primary and redundant chains exist on the satellite with the possibility also of a mixed configuration. Calibration between redundant and nominal units slightly change these values, but for the approximate comparison to other methodologies performed in this section only the active nominal chain will be used. The tracking configuration in the receiver needs also to be clarified, since pilot and data components on the same frequency may differ up to 0.2 ns as confirmed by ground and in orbit measurements.

Ground satellite measurements with a navigation receiver

Ground satellite tests with the satellite connected to the receiver provide a valuable data set to verify the group delay. The measurements have several advantages such as high signal to noise ratio due to the short distance and therefore low noise, no antenna code/phase pattern delays in the transmitting/receiving antenna, no multipath and no ionospheric effects. However, unknown additional cable and splitter/combiner delays in the set up affect the receiver hardware delay term. As there is no ionospheric delay the IFB is computed directly by differentiation of measurements:

$$IFB_{1,2}(t) = \quad P_{2r}^s(t) - P_{1r}^s(t) = (GD_{r2} - GD_{r1}) + (GD_2^s - GD_1^s) \qquad [6.4]$$

High gain antenna and triple carrier combinations

Triple carrier combinations may be applied to compute GIOVE IFBs as proposed by [153], where starting from the triple-frequency geometry-free iono-free combination of code ranges [152],

$$\lambda_3^2 c_{12} + \lambda_1^2 c_{23} + \lambda_2^2 c_{31} = \lambda_3^2(P_1 - P_2) + \lambda_2^2(P_3 - P_1) + \lambda_1^2(P_2 - P_3) \qquad [6.5]$$

applied to the (E1, E5a, E5b) combination and assuming that the code bias between E5a and E5b is in the sub-nanosecond range,

$$
\begin{aligned}
c_{12} &= GD_{E1}^s - GD_{E5a}^s \\
c_{23} &= GD_{E5a}^s - GD_{E5b}^s \approx 0 \\
c_{31} &= GD_{E5b}^s - GD_{E1}^s \approx GD_{E5a}^s - GD_{E1}^s = -c_{12}
\end{aligned}
\qquad [6.6]
$$

resolves c_{12} in Equation 6.5 by reordering the terms:

$$
\begin{aligned}
\lambda_3^2 c_{12} - \lambda_2^2 c12 &= \lambda_3^2(P_1 - P_2) + \lambda_2^2(P_3 - P_1) + \lambda_1^2(P_2 - P_3) \\
c_{12}(\lambda_3^2 - \lambda_2^2) &= \lambda_3^2 P_1 - \lambda_3^2 P_2 + \lambda_2^2 P_3 - \lambda_2^2 P_1 + \lambda_1^2 P_2 - \lambda_1^2 P_3 \\
c_{12}(\lambda_3^2 - \lambda_2^2) &= \lambda_3^2 P_1 - \lambda_3^2 P_2 + \lambda_2^2 P_3 - \lambda_2^2 P_1 + \lambda_1^2 P_2 - \lambda_1^2 P_3 + \underbrace{\lambda_2^2 P_2 - \lambda_2^2 P_2}_{=0} \\
c_{12}(\lambda_3^2 - \lambda_2^2) &= P_1(\lambda_3^2 - \lambda_2^2) - P_2(\lambda_3^2 - \lambda_2^2) + P_3(\lambda_2^2 - \lambda_1^2) - P_2(\lambda_2^2 - \lambda_1^2) \\
c_{12}(\lambda_3^2 - \lambda_2^2) &= (P_1 - P_2)(\lambda_3^2 - \lambda_2^2) + (P_3 - P_2)(\lambda_2^2 - \lambda_1^2) \\
c_{12} &= (P_1 - P_2) + (P_3 - P_2)\frac{(\lambda_2^2 - \lambda_1^2)}{(\lambda_3^2 - \lambda_2^2)}
\end{aligned}
\qquad [6.7]
$$

Finally, recovering the previous nomenclature the final expression can be written as:

$$IFB_{1,2}(t) = (P_{1r}^s(t) - P_{2r}^s(t)) - (P_{2r}^s(t) - P_{3r}^s(t))\frac{\lambda_1^2 - \lambda_2^2}{\lambda_2^2 - \lambda_3^2} \qquad [6.8]$$

Fig. 6.13: Chilbolton, Weilheim and Westerbork high gain antennas

The problem of the amplification of the noise due to the term $\frac{\lambda_1^2-\lambda_2^2}{\lambda_2^2-\lambda_3^2} \approx 17$ derived from E5a and E5b proximity can be overcome by the use of a single high gain antenna (HGA) directed to the satellite. These measurements have the advantage of a high signal to noise ratio, no multipath and no receiver antenna patterns. Particularly relevant are the parallel measurements with the 25 meters diameter high gain antenna at Chilbolton (UK), a different high gain antenna (30 m) and setup at Weilheim (Germany), and an additional one at Westerbork (The Netherlands) and a smaller 5 meters antenna in Rome. The antennas are depicted in Figure 6.13. Details about the measurement set up for the antenna location to track GIOVE-B can be found in [57] for Chilbolton and [165] for Weilheim.

It has to be highlighted that the non-negligible receiver group delays have been dropped in the derivation of Equation 6.8. Disadvantage are the unknown additional hardware delays introduced by the antenna and cables which need to be calibrated.

Network adjustments

Network measurements with omnidirectional antennas have the advantage of a better knowledge (easier to be calibrated) of the antenna and cable delays, and the availability of a higher number of observations with continuous visibility. However, due to the lower antenna gain this type of measurements suffers from the disadvantage of directional code delays, higher noise (lower signal to noise ratio) and multipath.

The IFB for the satellite from network measurements can be obtained by resolving Equation 5.37 in a least squares adjustment by solving for ionosphere and IFB values as described in Section 5.4. This approach is used in GIOVE mission based on GESS stations. Since it suffers from the limited number of sensor stations to compute the ionosphere, an alternative methodology can be performed by using IGS-IONEX values to remove the ionosphere contribution in Equation 5.37 and solving only for the unknown satellite and receiver IFB as described in [27].

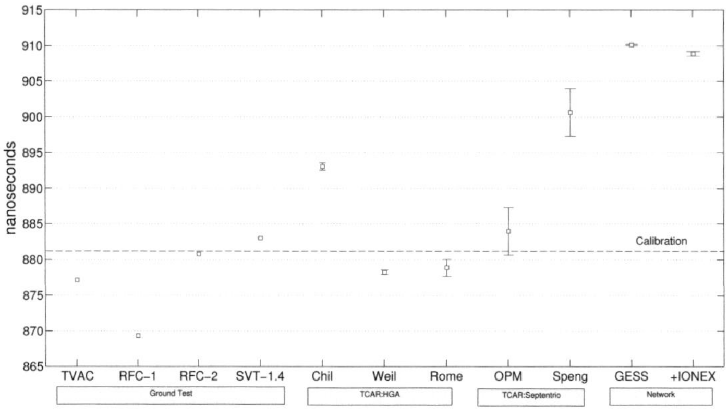

Fig. 6.14: GIOVE-B inter-frequency bias for L1B-E5b Pilot components at different tests

Summmary

As summary, Figure 6.14 compares the IFB between L1B-E5b Pilot components with the different methodologies. Calibrations are compared to ground tests, triple carrier combinations performed with the high gain antenna measurements and network estimations with ionosphere estimation or Ionex files usage. Independent values provided in [153] with a single receiver and two different choke ring antennas, a Space Engineering and anlernative antenna from Orban Microwave Products (OMP), are also included. Error bars represent the 1σ of the associated error.

Most of the different estimations agree within a few nanoseconds with the manufacturer calibrations. Ground tests (TVAC,RFC-2,SVT), high gain antenna (Weilheim and Rome) and triple carrier (OPM) are a few nanoseconds close to the ground calibration. Main differences are observed in the last three points (Speng, GESS and +IONEX) which have in common the use of omnidirectional Space Engineering antennas. A mean bias of around 30 ns is observed when comparing both satellites to the ground estimations in Table 6.3. As network estimations require to fix the ground receiver delays $(GD_{r2} - GD_{r1})$ to zero, the source of this bias seems to be an additional delay introduced between E1 and E5b frequencies by the Space Engineering omnidirectional antenna.

	GIOVE-A	GIOVE-B
Calibrations	888.45 ns	881.18 ns
GESS	922.40 ns	910.10 ns
Difference	33.95 ns	28.91 ns

Tab. 6.3: Inter-Frequency Biases in GIOVE-A and -B for L1B-E5b pilot from ground calibrations compared to estimations with GESS network.

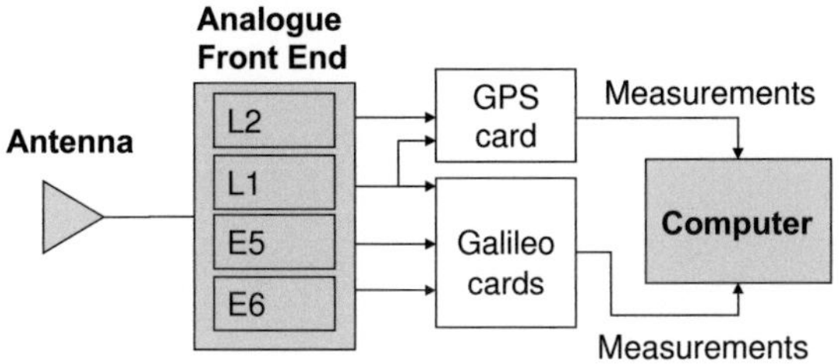

Fig. 6.15: GIOVE receiver high-level block diagram

As conclusion, absolute group delay values are difficult to be measured and practically impossible to be verified in orbit as is also acknowledged for GPS [162]. Only the IFB may be observed at payload level by different methods. Here, it has been demonstrated that it would be required to absolutely calibrate the ground test cables, the receiver and antenna of one or several stations to validate satellite calibrations. Nevertheless, the state of the art accuracy at BIPM for calibration of receiver and antenna is 5 ns [129] which is in-line with the inaccuracy observed in Figure 6.14. An advance in the station calibration techniques is required to improve the uncertainty in the group delay and differential group delay calibrations in orbit.

It has to be remarked that normal single frequency users are not affected by a common bias in all satellites broadcast group delay, as this bias is absorbed in the receiver time dt_r estimation. The important factor is the relative value between satellites and the evolution over time. In consequence, network adjustments are the better approach to follow the evolution of the IFBs.

6.4.2 Inter-system bias

On the receiver side the situation is slightly more complicated. The stations are driven by a single clock common to the GPS and the Galileo channels. The station antenna is also common for both systems. The GETR receiver manufacturer by Septentrio is the basic receiver within the GESS stations. A second receiver manufactured by Novatel is also available at 3 stations. Both receivers share a common design approach. They track GPS signals with a standard commercial board, the PolaRx2 in Septentrio and the Euro-3M in Novatel, while the Galileo signals are tracked by separate boards. The general design is depicted in Figure 6.15, detail designs are available in each manufacturer user manual [151, 117].

Since both boards are different, the same signal frequency follows a different path in each board and has different delays. Provided the same L1 signal is tracked by both boards the pseudoranges will present a bias. This technique can be used to roughly calibrate and align the channel delays between GPS-Galileo but antenna values are not aligned. As a consequence, different Group Delay biases are included in the apparent clock observed at the station for each system. Reusing the same nomenclature from Equation 5.34 :

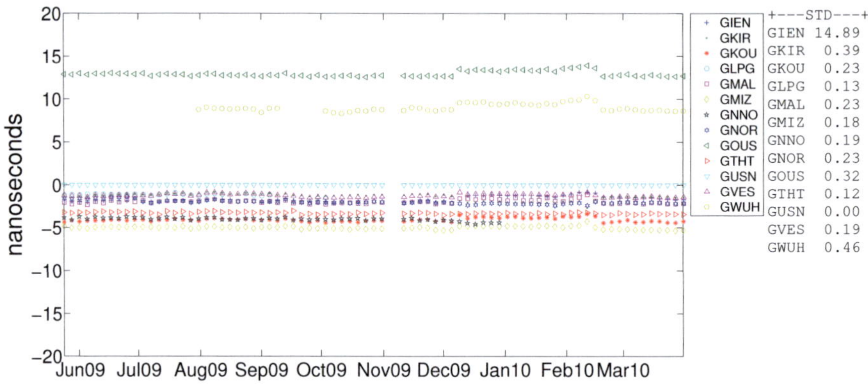

Fig. 6.16: Inter-system Bias in GESS network wrt. GUSN

$$dt_r'(GPS) = dt_r + K_{E1}GD_{E1r} - K_{E5ar}GD_{E5ar}$$
$$dt_r'(GAL) = dt_r + K_{L1}GD_{L1r} - K_{L2r}GD_{L2r} \qquad [6.9]$$

The difference between these two ionosphere-free clocks is called the inter-system bias (ISB),

$$ISB_r = (K_{E1}GD_{E1r} - K_{E5ar}GD_{E5ar}) - (K_{L1}GD_{L1r} - K_{L2r}GD_{L2r}) \qquad [6.10]$$

In GIOVE mission the observed station clock is normally referred to the GPS part of the station, since this is much better observable due to the many flying GPS satellites (as opposed to only two flying GIOVE satellites during the analysed periods).

The estimation of a station inter-system bias (ISB) is required when using observations from a combined GPS/Galileo station. For the reasons explained, an ISB is estimated for each station as a constant value by estimated arc. The stability of this estimated value is crucial as any instability would be absorbed by the satellite clocks.

The GIEN station has been selected as reference for inter-system bias investigations, since this station is also the reference for all clock estimations. The way to calibrate the actual inter-system bias of GIEN is based on estimates of the station IFBs obtained by processing geometry-free code and phase observables from the station. In this processing, the average of the estimated P1-P2 IFB of GPS satellites has been aligned with the mean of IFB values contained in the navigation message, as calculated by the Jet Propulsory Laboratory. The mean GPS BGD is around +8 ns corresponding to a mean IFB of around +5 ns (after multiplying by a L1/L2 frequency factor). The C1C-C7Q IFB of the GIOVE satellite has been fixed to the L1-E5 hardware delay calibrated by the satellite manufacturer in Table 6.2.

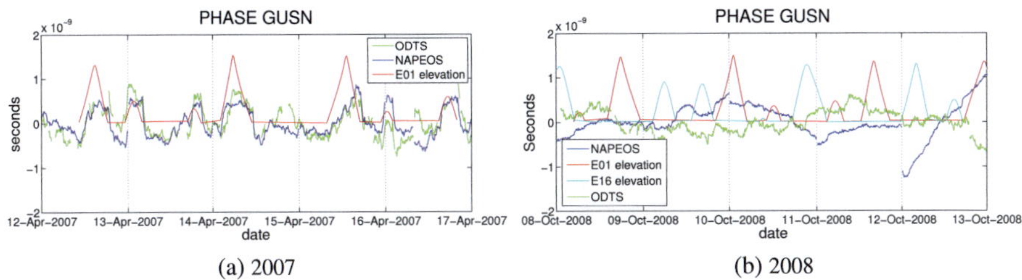

Fig. 6.17: GIOVE pass effect on the intersystem bias of GUSN station.

Figure 6.16 presents the ISB from June 2009 referred to GUSN in order to analyze the stability of the ISB. The standard deviation for the complete period is also included in the figure. All GESS stations have similar ISB values being grouped between 0 and -5 ns with respect to GUSN. Only two of the receivers GWUH and GOUS are out of family with slightly higher values. Several jumps are present in GIEN station due to problems at the station and subsequent updates of the receiver, as a consequence the standard deviation is not representative. The approach to fix the GUSN value to zero implies that any effect on the ISB at GUSN impact the rest of the stations, as observed from December-2009 till mid of February 2010 where all stations present a simultaneous step of 1 ns. The daily ISB for the stations is stable from 5-days to 5-days arcs used in GIOVE-M with an overall sigma of 0.2 ns as observed on the right table in Figure 6.16. It remains to clarify whether the subdaily values are also stable.

Subdaily stability

To check the subdaily ISB the clock estimations at GUSN station will be further analyzed. Figure 6.17 presents the station clock after drift removal for GUSN station during a GIOVE pass over in April 2007. The elevation of the satellite is over layed with a red line. ODTS solution has been performed with a 5 days long pass. The clock has been detrended for the 5 days ODTS solution with a order two polynomial, as a consequence, the first and last days can include higher error. An additional clock solution by an independent software (NAPEOS) has also been included. The NAPEOS clock is obtained from the central day of a 3 days arc fitting with a mixture of IGS and GIOVE-mission stations. The clock is detrended day introducing some jumps at the daily borders. The clock solution for GUSN with NAPEOS is referred to an IGS station driven by an H-maser while ODTS estimations are referred to GUSN being affected by the ISB at both stations.

For the beginning of the analyzed period from end 2006 till mid of 2007 in subfigure (a) the results are strongly affected by GIOVE visibility. The estimated phase offset for the station changes up to 1 ns peak-to-peak when GIOVE-A enters in visibility. For this period all eight

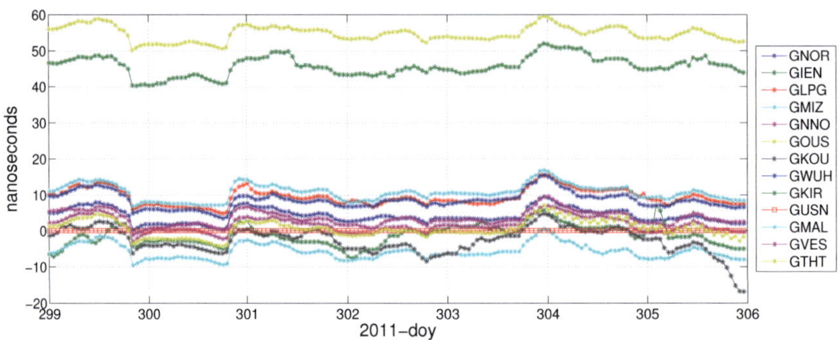

Fig. 6.18: Subdaily Inter-System Biases at GIOVE stations

channels are allocated to track GIOVE-A. As the time transfer is dominated by the GPS signals, the variation of the phase estimations at GUSN during 2007 could be attributed to variations in the ISB stability caused by an increasing temperature on the GPS channels due to GIOVE tracking on the eight allocated channels and the posterior deactivation of the same channels after the satellite leaves visibility. The effect is less evident from 2008 after the improvement of the cooling capabilities by Septentrio and the later semi-continuous tracking of GIOVE-A/-B signals. Nevertheless, hardware delay instabilities could be still present on the Galileo measurements and therefore in the ISB.

The ISB is typically estimated as a daily value by the POD adjustment, however the separated iono-processing in charge of computing the total electron content and group delay values allows to define a lower time interval estimation (methodology in Section 5.5). A higher variability of the group delay at the station is obtained from the ionosphere processing. One station has to be used as reference, in this case the GUSN value has been assumed as zero and all the other stations refer to it. As a consequence, any ISB in this station will affect all the others as clearly observed in Figure 6.18 where all stations follow a similar trend with small variations. It can be concluded that the ISB stability is a constant value with sub-daily variations.

6.5 One-way carrier phase time transfer

In previous Section 6.3 the limit of geodetic time transfer has been quantified as 1-2 E-12 with 0.1 to 1 ns accuracy, based on a large number (>100) of sensor stations to compute the satellites and orbit products. This stability is poorer than the expected PHM performance and some of the best performing RAFS. One of the main objective of GIOVE mission was to characterize the new AFS performance in space. A new methodology was required to validate the signal clock stability.

Clock estimations by POD are performed normally at 5 or 15 minutes intervals from which it is possible to extract the on board clock performance for this time interval or at higher τ values by integration. The question remains how the clock is performing below 5 min. Short term stability is particularly relevant for the receiver tracking noise and to derive the interpolation error in nominal 5 min clock products. In the last years, CODE has also started to produce clock information at 5 - 30 seconds in order to decrease the interpolation error for PPP applications. However, this product presents some drawbacks, since no other center provide data, the derived files are quite large and require more processing time and memory. Additionally, these estimations are not available for new GNSS constellations not formally included in the IGS processing as GIOVE, Galileo or COMPASS. Another method is required to check short term performance.

A new methodology is proposed here after to complement IGS limitations and fulfill GIOVE mission objectives. The use of one-way *code* measurements to characterize the short term behaviour of satellite clocks was already proposed in 1984 shortly after the GPS constellation deployment [80]. In those early days the code was the base to characterize space and ground clocks. The use of carrier phase for time transfer applications was proposed 15 years later by NIST [89, 90] and finally used officially to characterize the satellite clocks by USNO [119].

In this section the early proposed methodology by [80] is reviewed to use one-way carrier *phase* from one station with high sampling rate (e.g. 1 second) to characterize the satellite clocks and validate POD solutions. It will be demonstrated how this new proposed method developed within this dissertation equally applies for the characterization of ground and satellite clocks and allows a detailed analysis of the noise affecting the GNSS navigation signals.

6.5.1 Mathematical model

In order to compute the Allan variance [3] is used :

$$\sigma_y^2(\tau) = \tfrac{1}{2} \left\langle (\Delta y)^2 \right\rangle \qquad [6.11]$$

where y is the instantaneous fractional frequency for the particular τ interval and Δy is the difference between consecutive $y_k(\tau)$ estimations. Consequently, Equation 6.11 can be rewritten as:

$$\sigma_y^2(\tau) = \frac{1}{2(M-1)} \sum_{i=1}^{M-1} (y_{i+1} - y_i)^2 \qquad [6.12]$$

The instantaneous fractional frequency y_i is obtained as the derivative of the phase for the time interval τ. In this case ϕ is the carrier phase measurement as observed by the receiver tracking PLL.

$$y_i = \frac{dt^s(t_k + \tau) - dt^s(t_k)}{\tau} \qquad [6.13]$$

As a consequence, Equation 6.12 can also be expressed as function of the satellite clock phase offset dt^s as

$$\sigma_y^2(\tau) = \frac{1}{2(N-2)\tau^2} \sum_{i=1}^{N-2} (dt_{i+2}^s - 2dt_{i+1}^s + dt_i^s)^2 \qquad [6.14]$$

In the traditional POD based methods, explained in the previous chapter, dt^s is computed grouped with the group delays. Here the novelty will be to use directly the carrier phase measurement by assuming $dt^s = \phi$ and avoid the demanding POD computation.

Effect	meters
Sagnac effect	23.000
Orbit eccentricity	15.000
Space curvature	0.018
Shapiro	0.020
Phase wind-up	0.120
Tropospheric delay	2.230
Troposphere curvature	0.030
GPS satellite PCO	2.700
GPS satellite PCV	0.010
Receiver PCO	0.120
Receiver PCV	0.020
Satellite DCB	0.200
Receiver DCB	14.000

Tab. 6.4: Indicative magnitude of the different errors on GNSS ranges

As presented in Equation 5.27, besides the clock phase dt^s the carrier phase (ϕ) measurement performed by the receiver includes additional effects, such as the satellite to receiver dynamics, troposphere, periodic relativistic correction, phase wind-up, etc. Some of these variations can be accurately removed by models, such as the Sagnac and relativity terms, whereas others can only be partially removed since they are based on empirical models with some error associated, such as troposphere and antenna models. The magnitude of these delays is covered in Table 6.4.

Once the empirical contributions (PCV and PCO for receiver and satellite, troposphere slant delay, relativistic terms, phase wind-up and Sagnac effect) are removed from Equation 5.27, the new Equation 6.15 still includes the geometry, ambiguities, group delays, ionosphere and multipath sources.

$$
\begin{aligned}
\phi_{rk}^{s}(tor) = \ & |\bar{X}_r(tor) - \bar{X}^s(tot)| + \\
& + dt_r(tor) - dt^s(tot) \\
& + GD_{rk}(tor) - GD_k^s(tot) \\
& - 40.3 \frac{STEC_r^s(tor)}{f_k^2} \\
& + \lambda_k N_{rk}^s(t_0) + \lambda_k \left[\phi_{rk}(t_0) - \phi_k^s(t_0 - \tau_r^s(t_0)) \right] \\
& + PLL_{rk}^s(tor) + m_{rk}^s(tor) + i_{rk}(tor)
\end{aligned}
\qquad [6.15]
$$

Part of these contributions can be further reduced or eliminated: the geometry can be computed by using receiver X_r and satellite X^s coordinates, the accuracy of the latest depending on the source of the data used to perform the orbit determination e.g. radar (Km), S-Band tracking (m) or L-Band (cm); receiver clock instability $\sigma_{dt_r}^2$ can be brought below the satellite $\sigma_{dt^s}^2$ level by using a better AFS connected to the receiver (e.g. H-maser) or by correcting the clock by PPP estimation of dt_r; multipath m_{rk}^s can be significantly reduced by setting higher elevation masking angles (e.g. $> 30\ ^o$) or by using directive high gain antennas which in addition reduce PLL noise level and possible interferences $\sigma_{i_{rk}}^2$.

Finally, constant terms can be dropped since Equation 6.12 is based on the derivative of the phase.

$$
\begin{aligned}
\phi_{rk}^{s}(tor) = \ & -dt^s(tot) + GD_{rk}(tor) - GD_k^s(tot) \\
& - 40.3 \frac{STEC_r^s(tor)}{f_k^2} \\
& + PLL_{rk}^s(tor)
\end{aligned}
\qquad [6.16]
$$

It is possible to use ionosphere-free combinations to eliminate the first order ionosphere contribution but increasing significantly the PLL noise. Nevertheless, the ionosphere contribution does not affect the short term Allan variance for $\tau < 300$ sec since it is a slowly changing atmospheric effect. Only the group delays cannot be eliminated. Nevertheless, due to its thermal origin, they are only expected to contribute over 1000 seconds not affecting the objective of this methodology.

Finally, Figure 6.19 shows the schematic description of the elements involved in the methodology. It is possible to use only the omnidirectional antenna, however the inclusion of a high gain antenna provides lower PLL noise and render possible to decrease the $\sigma_y(1)$ from 2E-12 down to 1E-13 if higher stability is required, as is the case for PHM clocks.

6.5.2 Model implementation

The difficulty in the model implementation consists in the accurate removal of the empirical contributions. A straight forward method to correct the deterministic contributions is to correct the measurements by using the pre-processing of a Precise Point Positioning software. Once corrected with a priori models, the carrier phase may be directly used if a more accurate fre-

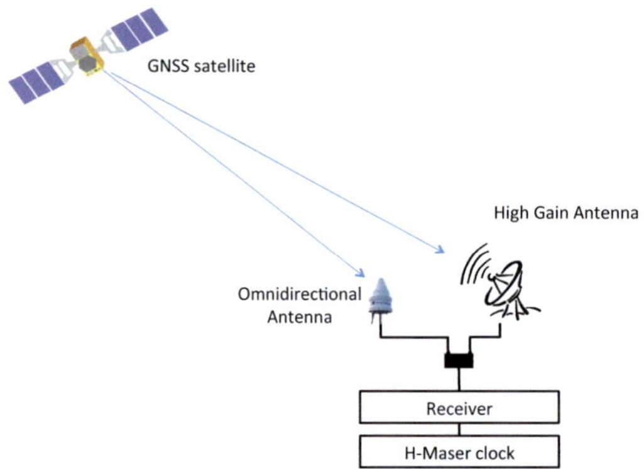

Fig. 6.19: OWCP schematic description

quency source is employed at the station as external frequency reference source (e.g.H-maser). Otherwise, the carrier phase residuals after computing the position can be used once corrected for receiver clock and troposphere contributions.

Several PPP consolidated software packages exist which might be used to correct the deterministic behaviour of ϕ. Unfortunately the currently available software present some limitations in order to perform this task, namely :

- processing of orbit (sp3) and clock products (clk) different than IGS is not always possible.

- processing of 1Hz observations is not always allowed. Data are usually down-sampled to 30 seconds.

- processing of Galileo frequencies and satellites is not implemented in any standard package.

- processing of Rinex 3.0 is not possible in any standard package.

- intermediate outputs (phase residuals) are not provided.

- source code is not available for modifications.

At the time of writing this manuscript no PPP software was freely available matching the required characteristics. The development of an ad-hoc new software has been performed based on [85]. This software has been validated against NRCAN solutions and final coordinates for a set of reference IGS stations. An outline of the different modules and corrections is available in Figure 6.20. Further details about the complete PPP software implementation can be found in [113].

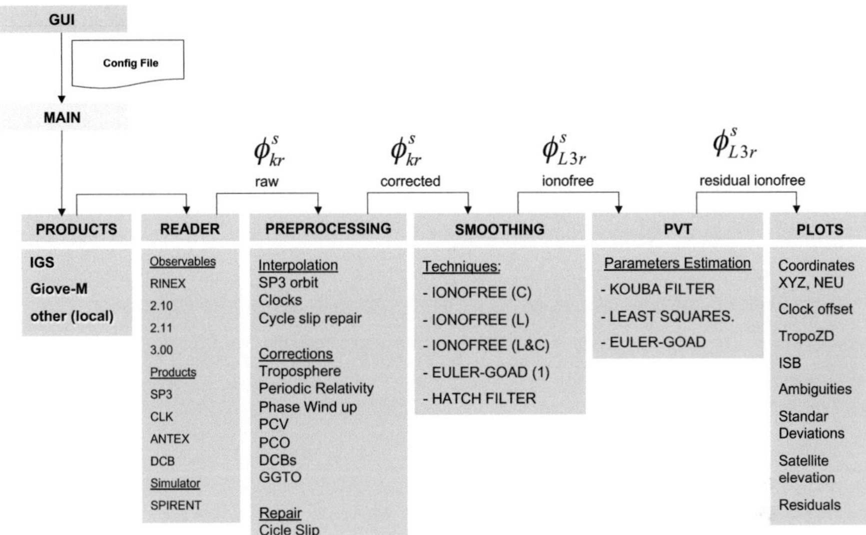

Fig. 6.20: OWCP PPP correction of phase observations

The *raw* carrier phase measurements ϕ_{kr}^{s} are read from the Rinex files and passed to a pre-processing module which corrects and repairs the observations for cycle slips. The *corrected* phase measurements are passed to the smoothing module which creates the basic ionosphere-free observables before entering the least squares adjustment. Here, the *corrected* and *ionosphere-free* carrier phase measurements after pre-processing are directly used with Equation 6.11 to compute the Allan deviation.

6.5.3 Model validation using GPS satellites

As input data to validate the model the observation files for a period of 5 days, covering the days 138 till 143 of the year of 2007, from two stations (GIEN and GUSN) connected to H-maser frequency standards and products (orbits, clocks, station coordinates) as estimated by GIOVE mission have been used.

Figure 6.21 presents the results for the complete GPS constellation. Stability obtained by this one-way carrier phase (OWCP) technique from 1 till 300 seconds are shown together with the stability derived from POD clocks obtained by IGS from 30 seconds and by GIOVE mission from 300 seconds. Detailed results are presented for each GPS Block clocks and signals, showing good agreement with 30 seconds IGS results. POD results are based on the ionosphere-free linear combination while the OWCP presents three different solutions, one for each single frequency and the other for the ionosphere-free combination.

As a general remark, OWCP solutions based on single frequencies (L1, L2) and the ionosphere-free linear combination match perfectly for integration times over 30 sec. Below 30 seconds, the ionosphere-free measurement noise is higher due to the combination of both signals. L1 noise is $2 \times 10^{-12}\tau^{-1}$ as expected for the carrier phase noise (PLL) with 40-50 C/N0 (see Figure 6.1). L2 does not present a τ^{-1} slope associated with white phase noise, indicating some type of coloured noise as also demonstrated by zero base line analysis in other publications. The usage of semi-codeless techniques to track L2 could be the origin of the coloured type of noise.

Each satellite clock type in Figure 6.21 has a different signature. The first τ values are dominated by the DLL tracking noise of the carrier phase but afterwards each clock technology is clearly identified :

- **Block IIA** frequency stability in cesium mode is composed of a quartz crystal oscillator at short term (0-10 sec) locked to the cesium frequency at longer intervals. PLL noise is overcome from 2-10 seconds by the internal crystal oscillator until it reaches the cesium frequency modulation white noise at $1.2 \times 10^{-11}\tau^{-1/2}$ around [30-100] sec. Transition from the crystal to the cesium depends on the satellite.

- **Block II-A** frequency stability in Rubidium mode is driven by the free running rubidium clock signal. Clock noise $5 \times 10^{-12}\tau^{-1/2}$ is higher than the PLL tracking noise. The carrier phase WPM noise is observed only for the ionosphere-free clock between 1-2 sec, afterwards the rubidium noise becomes dominant.

- **Block II-R** frequency stability is driven by the time keeping system. In this satellite type a 10.23 MHz digitally controlled VCXO is linked to the RAFS by a software controlled loop to produce a navigation signal with the timing accuracy of the RAFS. The noise presents several transitions : PLL noise in the [0-5] seconds interval with $2 \times 10^{-12}\tau^{-1}$, VCXO in [5-100], WPM in [100-3000] until finally from 3000 seconds the specifications of the PerkinElmer RAFS-IIR rubidium standard are valid ($3 \times 10^{-12}\tau^{-1/2} + 5 \times 10^{-14}$). This behaviour is in line with the description of the TKS+RAFS described in Section 4.3.

In the short term OWCP and IGS solutions overlap on the [30,300] sec interval with an excellent agreement which fully validates this technique. In general, all GPS clock families behave as expected from the description of the atomic frequency standard and frequency distribution unit performed in the previous Chapter 4. The only unexpected behaviour is obtained for cesium families with a better behaviour at short term. Normally the specified AFS noise is also assumed for short intervals in radio frequency constellation simulators or performance studies. This approach leads to wrong assumptions for cesium and Block-IIA rubidium clocks.

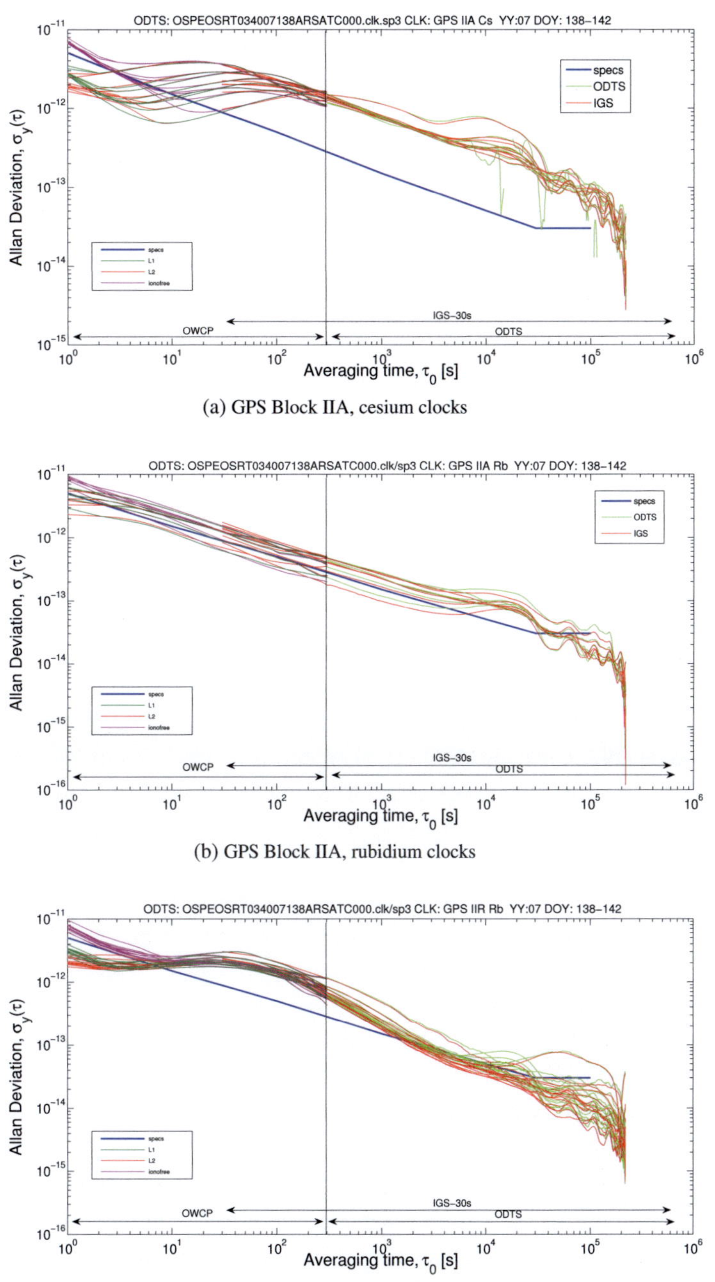

(a) GPS Block IIA, cesium clocks

(b) GPS Block IIA, rubidium clocks

(c) GPS Block IIR, rubidium clocks + Time Keeping System

Fig. 6.21: GPS satellites: one-way carrier phase (from 1 to 300 seconds) versus precise orbit determination (above 30 seconds with IGS and above 300 seconds with ODTS)

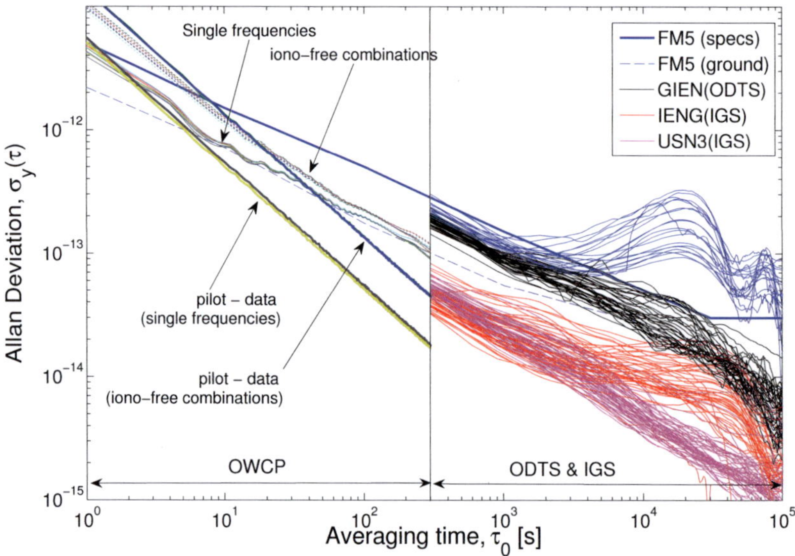

Fig. 6.22: GIOVE-A: one-way carrier phase (1-300) versus precise orbit dertermination (>300 sec)

6.5.4 Model validation using GIOVE satellites

Results for GIOVE-A are provided in Figure 6.22 in the 1-300 seconds interval with OWCP methodology and from 300 seconds with POD estimations from GIOVE mission.

In the [1-5] seconds interval the PLL noise $5 \times 10^{-12}\tau^{-1}$ is dominant until from 5 seconds on the real RAFS noise $2 \times 10^{-12}\tau^{-1}$ is reached. The same noise behaviour as on ground test is observed in the interval [5-300]. From 300 seconds based on POD estimations the measurement noise is higher. The right figure presents additionally the OWCP results for ionosphere-free and pilot minus data combinations. The pilot minus data combination eliminates all common errors (including multipath) allowing for a pure characterization of the carrier phase tracking error. It confirms that the $5 \times 10^{-12}\tau^{-1}$ noise for the single frequency signals and the $2 \times 10^{-11}\tau^{-1}$ noise for ionosphere-free combinations are associated to the PLL tracking.

The OWCP and ground tests for this RAFS present a good agreement. However, OWCP and POD results do not agree at 300 seconds. It seems as the clocks derived from POD are noisier than expected. This hypothesis is further analysed with the more accurate signal provided by the PHM on GIOVE-B. The POD results in Figure 6.23 have been obtained with 6 months of continuous data based on E1b-E5b ionosphere-free linear combination and GUSN as reference station. The effect of a harmonic of 0.5 ns amplitude with a period equal to the orbital period of the satellite has also been included.

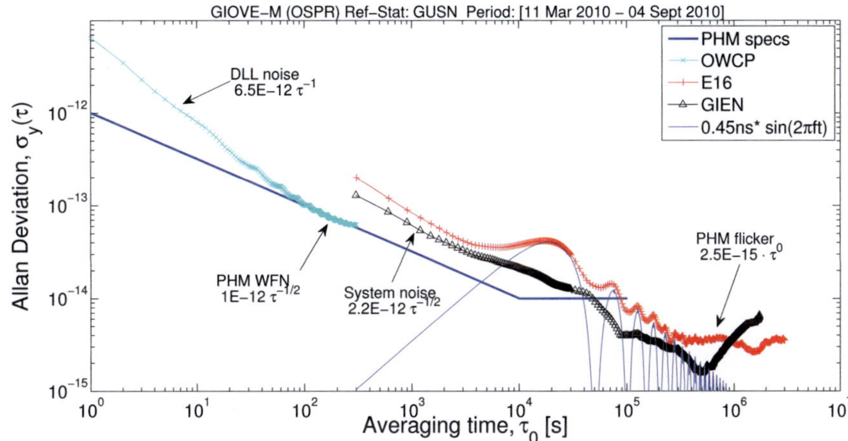

Fig. 6.23: GIOVE-B: one-way carrier phase (1-300 sec) versus precise orbit dertermination (>300 sec)

From one second the PLL noise $6.5 \times 10^{-11}\tau^{-1}$ is dominant until around 100 seconds where it reaches the PHM noise specifications $1 \times 10^{-12}\tau^{-1/2}$ which follows for the rest of the interval till 300 seconds. POD estimation from 300 seconds on follows a $2.5 \times 10^{12}\tau^{-1/2}$ slope until the flicker floor 2.5×10^{-15} of the clock is reached at 3.5 days. The previous Section 6.3.3 has identified the noise of stations connected to a H-maser frequency standard as good indication of the measurement system noise. The system noise represented by the noise for GIEN station presents a similar magnitude $2.2 \times 10^{12}\tau^{-1/2}$ which identfied the results obtained by POD as system noise.

An excellent agreement between OWCP, POD and the clock specifications is observed for GPS clocks. For GIOVE satellites the situation is different. Excellent agreement exists between OWCP and ground tests, whereas POD(ODTS) results are limited by the system noise being above the expected values for RAFS and PHM standards. A harmonic component can be observed over the POD system noise. This issue deserves further attention, being analysed in Chapter 7.5.

As a conclusion, the good agreement also for GIOVE validates the OWCP methodology. Furthermore, it can be initially concluded that GIOVE frequency standards have the same stability (WFN) in-orbit as on-ground and more important that no other unit in the payload chain introduces a higher noise than the AFS. To finally confirm this hypothesis it is necessary to compare these results against similar results from ground test.

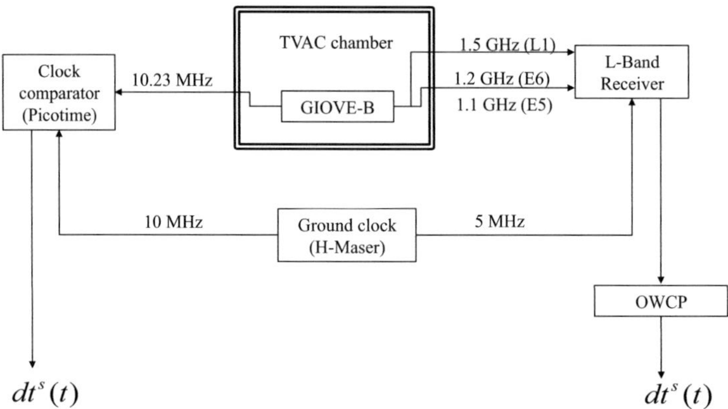

$$dt^s(t) \qquad dt^s(t)$$

Fig. 6.24: GIOVE-B clock stability test during TVAC

6.5.5 Comparison against ground tests

Ground tests are a golden reference to assess the performance of the orbiting AFS. On ground the clocks are tested before delivery to the satellite manufacturer to demonstrate the compliance against the specifications. The satellite manufacturer integrates the AFS in the satellite and they are further tested at payload and satellite level. Relevant tests for the AFS performance need to be performed in vacuum. The thermal vacuum test, where the satellite is exposed to an environment similar to space, is the most relevant test for performance verification. Test significance is however always limited in time and conditions. RAFS technology may need several weeks of operation before meeting the expected performance, a long observation time is required to verify the long τ values for any clock technology and different illumination conditions are expected in the satellite leading to different thermal profiles along the transmission chain. Unfortunately, due to cost and schedule limitations, it is not possible to test the satellite for longer than the strictly necessary time and tests are limited by the thermal chamber limitations.

This section presents the results and conclusions of the clock stability test performed for GIOVE-B during the satellite level Thermal Vacuum (TVAC) test phase, at TAS-I premises in Rome in August 2007. The test set up is depicted in Figure 6.24

One of the objectives of the test was to provide additional measurement data such that they can characterise the signal clock stability. This term refers to the stability of the signal measured by the Septentrio breadboard receiver when connected directly to the test output port of the Payload transmission chain. It is therefore representative of the true clock stability, degraded by the effects of the Payload transmission chain, the receiver chain and the measurement set-up.

Carrier phase measurements are obtained for each signal. Since the signal is not emitted, no propagation losses affect the signal. A high power signal is provided to the receiver which needs to be attenuated to dB values acceptable to the receiver (e.g. 70dB-Hz). The carrier phase data are characterized by a high signal to noise ratio and, therefore, by a significant lower noise than typical measurements with an omnidirectional antenna. The distance is fixed during the test, the temperature is stable at cold and hot levels and no propagation delays are introduced by the environment as the signal is not emitted. Consequently the ground phase measurements do not require the corrections applied in Section 6.5.2 to the in-orbit measurements. Carrier phase measurements obtained by the GETR are mainly driven by the clock noise and can be used directly to compute the stability of the satellite clock against the PHM connected to the receiver.

It is possible to derive the new propagation equation for the phase measurements by eliminating in Equation 5.27 the terms which are no longer present, such as: geometry terms, antenna patterns, troposphere, relativity, Sagnac, multipath and interference. Only clocks, group delay, ambiguities and PLL tracking noise remain in the phase measurements in the new propagation Equation 6.17:

$$
\begin{aligned}
\phi_k^s(tor) = \ & dt_r(tor) - dt^s(tot) \\
& + GD_{rk}(tor) - GD_k^s(tot) \\
& + \lambda_k N_{rk}^s(t_0) + \lambda_k \left[\phi_{rk}(tor_0) - \phi_k^s(tor_0 - \tau_r^s) \right] \\
& + PLL_{rk}^s(tor)
\end{aligned}
\qquad [6.17]
$$

Since the Allan deviation is based on the derivative of the phase also the constant ambiguity terms disappear, group delay variations GD and PLL noise remaining as the only terms disturbing the clock dt stability analysis.

$$
\begin{aligned}
\phi_k^s(tor) = \ & dt_r(tor) - dt^s(tot) \\
& + GD_{rk}(tor) - GD_k^s(tot) \\
& + PLL_{rk}^s(tor)
\end{aligned}
\qquad [6.18]
$$

From these phase measurements ϕ it is possible to extract the fractional frequency deviation and the Allan deviation by adopting Equations 6.11 or 6.13 used for the one-way carrier phase methodology.

Besides the raw carrier phase also two linear combinations are created. First, the pilot minus data combination on the same signal cancels all common contributions leaving only the PLL noise introduced by the receiver. Group delay variations in pilot and data signal are also expected to be almost the same, since they follow the same path and share the same frequency dependent delays.

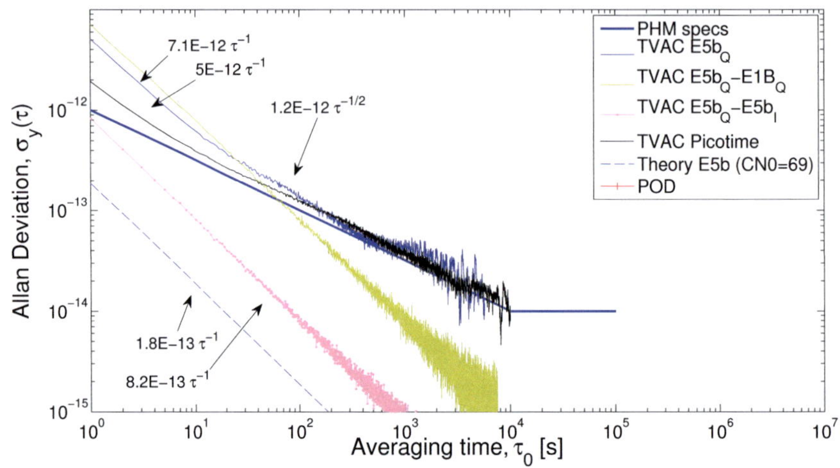

Fig. 6.25: GIOVE-B clock stability measurement during thermal vacuum test

$$\phi^s_{rE5bQ}(tor) - \phi^s_{rE5bI}(tor) = \left(GD_{rE5bQ}(tor) - GD^s_{E5bQ}(tot) \right) - \left(GD_{rE5bI}(tor) - GD^s_{E5bI}(tot) \right)$$
$$+ PLL^s_{rE5bQ}(tor) - PLL^s_{rE5bI}(tor)$$
$$\approx PLL^s_{rE5bQ}(tor) - PLL^s_{rE5bI}(tor)$$

$$[6.19]$$

Second, an inter-frequency combination (E5b- E1B) is also created; since no ionosphere is affecting the measurements it is possible to subtract directly two different signal measurements. Common receiver and satellite clock errors should be eliminated leaving the different hardware delay variations and thermal noise. This combination should provide equivalent results than Equation in terms of Allan deviation at constant temperature, where no variations in the group delays are expected.

$$\phi^s_{rE5bQ}(tor) - \phi^s_{rE1BQ}(tor) = \left(GD_{rE5bQ}(tor) - GD^s_{E5bQ}(tot) \right) - \left(GD_{rE1BQ}(tor) - GD^s_{E1BQ}(tot) \right)$$
$$+ PLL^s_{rE5bQ}(tor) - PLL^s_{rE1BQ}(tor)$$

$$[6.20]$$

Figure 6.25 presents the TVAC results during the hot level at constant higher temperature versus the orbit results. Several interesting conclusions can be extracted.

The Allan deviation derived from the single carrier phase signal $E5b_Q$ follows a $5.1 \times 10^{-12}\tau^{-1}$ slope until it converges towards the picotime results at 200 seconds. Picotime results follow a $1.2 \times 10^{-12}\tau^{-1/2}$ slope which are the specifications of the ground PHM used as reference in the test. The reference PHM used by picotime and receiver was slightly less accurate than the space PHM ($10^{-12}\tau^{-1/2}$).

Interesting is the short term noise below 200 seconds analyzed hereafter. Since the short term noise follows a τ^{-1} slope only 1 second τ values are retained. Results for combinations are divided by $\sqrt{2}$. The value of 5E-12 at one second for $E5b_Q$ is quite surprising since the expected value from theory is 1.8E-13 following Equation 6.2, what is one order of magnitude different. The combination pilot minus data component for E5b ($E5b_Q - E5b_I$) is more in line with the expected value 8.2E-13/$\sqrt{2}$=5.8E-13. The last inter-frequency combination between $E5b_Q - E1B_Q$ provides also unexpected results. The noise level for this signal is 7.14E-12/$\sqrt{(2)}$=5E-12 which is the observed noise for the single E5b carrier instead of providing similar values to the pilot minus data combination.

The explanation of these unexpected results are on the in-orbit measurements. Similar noise as measured in orbit by OWCP in Figure 6.23 is also observed in the ground test. However, in orbit result is in line with the expected value for C/N0=40 dB-Hz. In case the measurements with the maximum signal to noise ratio are used (50 dB-Hz), the noise observed in orbit decreases to 1.5E-12. Since the orbit measurements do not reflect the higher 5E-12 value. it suggests that the splitter, attenuators and cables used in the TVAC test were introducing some additional white phase noise.

Finally, it can be concluded that the signal clock noise for the GIOVE-B satellite is due to the physical clock. No further stochastic noise is introduced by other payload components, such as the frequency distribution unit. Additionally, it also demonstrates how the proposed OWCP technique can be easily adapted to ground test to verify the satellite performance.

6.6 Conclusions

In this chapter has been presented how the precision of Geodetic time transfer can be derived in a step wise approach.

First, Section 6.2 has identified an important conclusion, for a typical omnidirectional antenna the theoretical one-way time transfer accuracy limit is 100 ps using code and 1 ps by phase. As more than one satellite is normally in view this accuracy could be increased by averaging.

Second, the internal consistency can be validated from the adjustment residuals where code and phase residuals are expected to be randomly distributed and also within the theoretical noise level, by repeatability of results and by checking the expected precision of a reference standard (in this case the clock noise of stations with H-masers). The absolute accuracy can be validated by the reproducibility by independent processings, using the same or independent data, and by independent techniques, such as SLR for orbits and TWSTFT for clocks.

The practical example of the GIOVE+GPS satellites estimation with the use of the GIOVE-M network of 13 stations has been reviewed and compared with IGS methodology. A summary of the comparison is provided in Table 6.5 against the state of the art in POD achieved by IGS.

	GPS		GIOVE	
	Radial	Clock	Radial	Clock
IGS	0.02	0.02		
GGSP	0.02	0.03	0.31	0.43
ODTS	0.10	0.25	0.32	0.30

Tab. 6.5: Radial versus clock precision (1 σ) for GPS and GIOVE satellites in nanoseconds [ns]

GPS satellites achieve for clock products 0.07 ns (rms) accuracy and 0.02 ns (1σ) precision for orbit and radial components. The precision of GIOVE estimations by GGSP and GIOVE-M is 0.3 ns (1σ) with an estimated accuracy of 0.5 ns (rms), which is in line with SLR residuals. Better performance for GPS than for GIOVE satellites is expected due to the higher number of sensor stations. This hypothesis has been confirmed by analyzing the GPS estimation dependency on the number of sensor stations and by the fact that longer estimation arcs of 5 days increase the accuracy of the orbit estimation with respect to the nominal 1-3 days arcs used in IGS.

The accuracy limit of the geodetic time transfer is expected to be 0.1 ns. It has been reviewed with respect to TWTSFT time transfer. Both have been demonstrated to be consistent at 2 ns level, TWTFST being noisier at short interval times but converging to POD at longer integration times.

Validation of group delays estimations has been demonstrated as a challenging task. GIOVE satellites and new frequency combinations allow for a deeper insight into these values. IFB for GIOVE satellites have been reviewed with uncertainties below 5 nanosecond level with respect to the satellite calibrated values. On the stability side, satellite group delays have been demonstrated to not be as stable as assumed, with absolute variations due to changes at the stations and seasonal and sub-daily variations. Since the station group delays (ISB and IFB) are assumed to be constant during the estimation arc, any variation will be propagated into the reference time scale and satellite clocks.

The average behaviour of the best H-maser estimations can be considered to be the limit of the geodetic time transfer. The analysis of stations with active H-maser confirms that the stability of geodetic time transfer is $1\text{E-}12\tau^{-1/2}$ and therefore still noisier than the signal provided by the H-maser ($1\text{E-}13\tau^{-1/2}$) and at the limit of new Galileo PHM and Block-IIF frequency standards ($1\text{E-}12\tau^{-1/2}$).

Due to the limited number of stations and group delay instabilities at the stations, the achieved geodetic time transfer stability by GIOVE mission ($2.2\text{E-}12\tau^{-1/2}$) has been confirmed to be noisier than the PHM and best performing RAFS in GIOVE satellites. As explained in Chapter 2, no European atomic clock had previously been launched into space and the verification of its in-orbit performance was the second main objective of the GIOVE mission. Another methodology was required to verify the in-orbit performance of GIOVE clocks.

A new methodology has been proposed within this thesis using carrier phase measurements obtained at a station connected to a H-maser as frequency source. The new proposed methodology has been described and implemented in a dedicated piece of software and then validated with GPS satellites by comparison against IGS results with an excellent agreement. For the first time, the short term behaviour below 300 seconds not covered by IGS final products has been characterized, allowing for, in combination with POD results, full characterization of GNSS clocks from 1 second on.

Once its suitability to characterize GNSS clocks was confirmed, it has been applied to GIOVE clocks. As a result, it has now been proved how the short term stability of RAFS and PHM are in line with the ground measurements, being even possible to identify the activated RAFS unit from the agreement. Additionally, the same methodology has subsequently been modified and successfully applied to satellite ground tests to validate the stability of the *signal clock* on ground.

These analyses have allowed the validation of this novel methodology, the first full characterization of GIOVE and GPS clocks, and the successful achievement of the second main objective of the GIOVE mission.

7 Harmonics in satellite clocks

7.1 Introduction

During the accuracy assessment performed in the previous Chapter 6.6 the harmonics in GNSS apparent clocks have been identified as an ambiguous effect difficult to be attributed to clock or orbit sources due to the coupling between both components. Once the difference between the ground AFS stability and POD stability has been understood it remains to explain the 'bump' in the Allan Deviation introduced by a harmonic of typically 0.5 amplitude in GIOVE PHM 'apparent' clock phase.

All GNSS signal clocks show a periodic fluctuation. Harmonics in GPS satellites are a well-known feature since the early estimations of GPS clocks [158]. The impact on the clock prediction was also early acknowledged, IGSMAIL-3057 already in 2000 suggested to the different analysis centers to include the harmonic in the prediction for ultrarapid products. Nevertheless, their characteristics and origin have been only recently characterized [150]. Amplitudes of several nanoseconds are reported for Block-IIA satellites while values lower than 0.2 ns are observed in Block-IIR. Amplitudes for GIOVE satellites are definitively larger with values in the order of 1 ns for GIOVE-A RAFS and 0.5 ns for GIOVE-B. This periodic fluctuation in phase seems to be always present with a period analogous to the orbital one ($\approx$14 hours). To understand if this is a real physical phenomenon in the 'signal clock' or whether is a residual of the POD in the 'apparent clock' several analysis steps may be performed.

First the harmonics must be confirmed by different independent estimations. Once the harmonic is confirmed not to be an artifact of a single processing, three possible causes can be identified:

1. Frequency variation originating in the atomic frequency standard (the pure physical clock) due to sensitivity to temperature.

2. Group delay variation (in the signal clock) originating from one or several payload units due to sensitivity to temperature.

3. Orbit residuals (in the apparent clock) due to an estimation error of the orbit. As covered in previous sections radial orbit errors and clock errors are close correlated in the POD process.

The next Chapter will analyze the impact of the prediction to the user in Section 8.6. In this chapter the route source of the harmonic in GNSS clocks is analyzed in order to identify the origin, which shall be understood towards the implementation of possible mitigation strategies at system or user level.

7.2 Confirmation by different SW estimations

In case the harmonic is a real feature in the signal, different independent estimations should observe the same harmonic with the same characteristics of period, phase and amplitude. This can be checked for GPS and GIOVE satellites with week 1509 processed by the GGSP and GIOVE-M. Figure 7.1 presents the clocks detrended day by day and the associated Allan deviation for each day. For GIOVE RAFS (E01) all analysis centers recover the same harmonic in phase and amplitude (1 ns), some differences are observed in amplitude for the last days which may be due to missing observations. For the selected GPS RAFS (G29, Block II-RM) the harmonic is the same in phase and amplitude (0.5 ns), only the ODTS (OSPE) presents a slightly higher amplitude as also observed in the Allan deviation. On the contrary, in the PHM (E16) case no homogeneity neither in phase nor in amplitude is observed between the analysis centers. Only for some periods there is a fair agreement. A ground station with a H-maser (GUSN) has been also selected to show that this effect is exclusive to satellite clocks. For this station it can be observed how ESOC and AIUB present similar results while the ODTS (OSPE) shows an additional noise introduced by the selection of GIEN as reference station. Nevertheless, no clear harmonic is identified for this station clock.

It can be concluded that since the same harmonics are observed for E01 and G29 by different software and networks of stations the effect must be associated to the satellite. For the case of E16 (PHM), as no agreement exists, it is not possible to extract a consolidated conclusion in this sense from this comparison.

7.3 Harmonic and temperature variations

7.3.1 Correlation with sun-beta angle

It has been mentioned that the harmonic amplitude changes over time. For interpretation of the dependency on temperature profiles it is helpful to study the evolution of the amplitude with respect to the sun-beta angle. This may give additional insight into the clock variations and clearly point to an oscillation in the group delay or clock thermal sensitivity.

The beta angle is defined as the angle between the orbit plane and the satellite-sun vector as observed in Figure 7.2. The beta angle may change between $\pm 90°$ depending of the orbital plane. At higher beta angles the yaw angle changes slowly while at low beta angles the yaw angle has a stronger variation to keep the panels pointing to the sun. As a consequence the sun

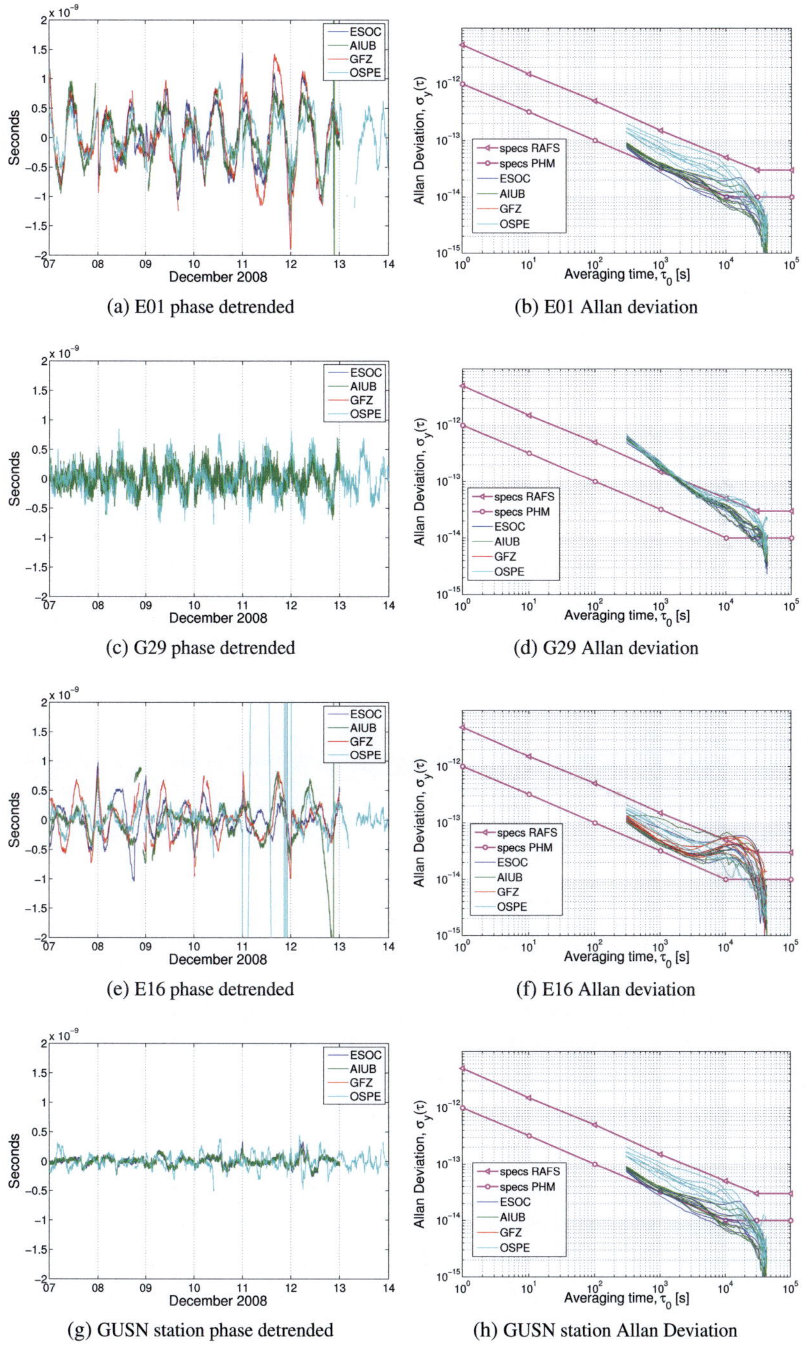

(a) E01 phase detrended

(b) E01 Allan deviation

(c) G29 phase detrended

(d) G29 Allan deviation

(e) E16 phase detrended

(f) E16 Allan deviation

(g) GUSN station phase detrended

(h) GUSN station Allan Deviation

Fig. 7.1: GGSP: GPS-Week 1509

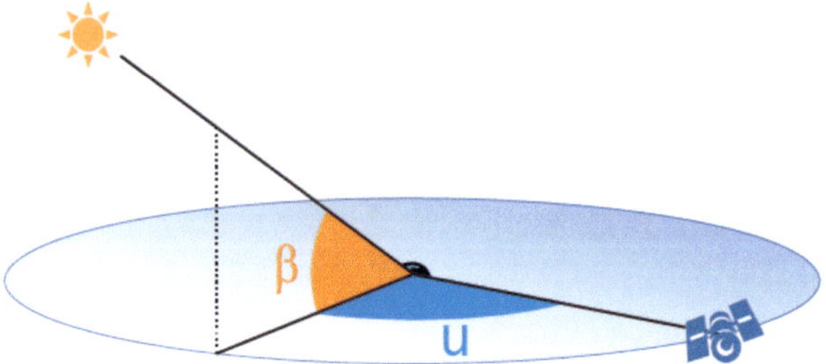

Fig. 7.2: Sun beta angle of the orbital plane and the attitude control rule the thermal enviroment of the satellite

illumination variation on the panels and thermal variation is higher. In case the harmonic source is linked to a temperature source, the phase peak to peak variations are expected to be lower at maximum beta-angle and higher at minimum beta-angle.

According to [106] the sun-beta angle of the plane containing the satellite at time t can be computed as :

$$\beta = \arcsin\left(\cos(\delta_\odot)\sin(i_0)sin(\Omega - \alpha_\odot) + sin(\delta_\odot)cos(i_0)\right) \qquad [7.1]$$

where :

- $\delta_\odot$ Declination of the Sun.

- $\alpha_\odot$ Right ascension of the Sun.

- i_0 Inclination of the satellite orbital plane.

- Ω Right ascension of the ascending Node (RAAN).

Values are related to the equator in the standard FK5 system, with respect to the standard equinox J2000.0. For the required accuracy the Sun position ($\delta_\odot$ and $\alpha_\odot$) can be estimated from tabulated values (see Chap.26 in [106]) and the satellite RAAN extracted from the navigation message i.e. the longitude of ascending node (at ephemeris reference epoch) minus the longitude of the vernal equinox at the ephemeris reference epoch.

All satellite orbits are subject to a drift in the parameter RAAN that is caused by natural orbit perturbations, particularly the first order secular drift rate due to the oblateness of the Earth. As a consequence, the RAAN drift rate of the GPS orbits is $-14.15°$ per year and that of the Galileo orbits $-9.01°$ per year [91]. Orbits in the same altitude and for the same inclination are all affected in the same way, so that the spacing between the orbital planes of one constellation is not affected. The constellations drift as a whole. For the analysis of the sun beta angle hereafter, the RAAN for each satellite has been obtained for a reference epoch in the middle of the analyzed period (03/09/2009). Some error should be expected in the sun-beta angle value from this approximation.

In order to obtain the amplitude of the 1st harmonic by day, a least squares fit (quadratic polynomial + 1/rev harmonic + 1/2rev harmonic) has been performed for each calendar day to each satellite with the following Equation 7.2:

$$x = a_0 + a_1 * t + a_2 * t^2 + a_3 \sin(2\pi f t) + a_4 \cos(2\pi f t) + a_5 \sin(\pi f t) + a_6 \cos(\pi f t) \qquad [7.2]$$

where f^{-1} is the period of the orbit (GPS = 12 h, GAL = 14 h) and the amplitude of the first harmonic is computed as:

$$A = \sqrt{a_3^2 + a_4^2}; \qquad [7.3]$$

The evolution of the harmonic amplitude for the different satellites may be observed in Figure 7.3 for a three year period. GIOVE satellites and a subset of GPS satellites, PRN-24 and PRN-32, of different blocks and clocks have been selected from different orbital planes. The left subplot depicts the temporal evolution, where each point in blue represents the amplitude for each day. A moving average of 10 days is also plotted as dark blue line. Unfortunately, it is affected by the numerous operations on GIOVE satellites and gaps in the processing, but it has been retained as for GPS satellites facilitates the interpretation of the results. The sun-beta angle is overlaid in red related to the second red axis. The right subplot presents the correlation of the amplitude for the first harmonic with respect to the sun beta angle computed with the raw estimations (red) and with filtered estimations after applying a $<5\sigma$ filter (green).

GIOVE-A RAFS (E01) shows a clear correlation with amplitudes of 1 ns at the highest beta angle (80°) raising up to 2 ns at the lowest beta angle. GIOVE-B PHM (E16) presents a low correlation with differences between higher and lower beta angle below 0.1 ns. Both GIOVE estimations have a large dispersion which can be attributed to the quality of the orbit.

Block IIA estimates evidence clear correlations for cesium and rubidium clocks. The selected PRN-24 (cesium) presents minimum values of 2.7 ns at maximum beta angles and 3.7 at minimum. The PRN-32 (Rb) also presents a clear correlation with 0.9 ns amplitude at maximum beta angle and 1.2 ns at minimum.

Block II-R driven by Rubidium clocks steered by the TKS appear to have a lower temperature dependency with a wider behaviour between the individual satellites. While some satellites

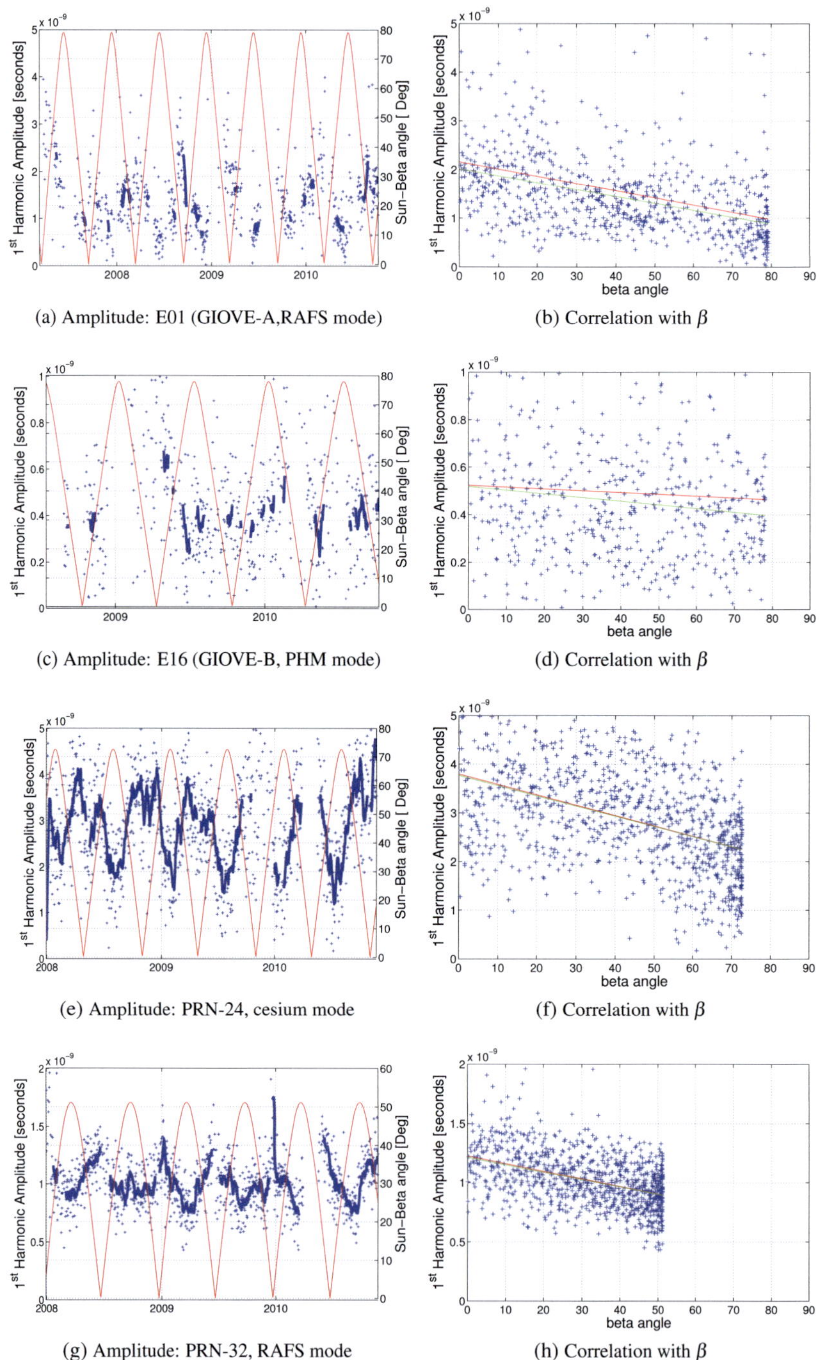

Fig. 7.3: 1st harmonic with respect to sun-beta angle (β) for different satellites

present a clear correlation as PRN-11 (SVN-46) with differences between maximum and minimum beta-angle of 0.2 ns amplitude, for others the correlation is low with differences lower than 0.05 ns as for PRN-17 (SVN-53) or PRN-14 (SVN-41).

It can be concluded that the source of the harmonics is due to thermal variations for the satellites presenting amplitude variations with a clear correlation with the sun beta-angle.

7.3.2 Analysis of temperature sensitivity of AFS

Physical clocks are sensitive to several environmental factors as humidity, magnetic field and temperature. Frequency standards on board of the satellite do not suffer from humidity in space vacuum. The magnetic field is well predicted and the unit is significantly isolated. Thermal control represents one of the main difficulties with considerable diametrically opposed outside temperatures, up to +100 °C in sun illuminated areas and down to -100 °C in shadow, and thermal requirements for the clock as low as $\pm 1^o$C at the physical clock [154]. GPS and Galileo satellites allow the outgassing of the platform internal gasses and use the space vacuum and radiators for a better thermal regulation, whereas GLONASS uses a pressurized gas system [14]. The next generation GLONASS-K satellites is expected to be based on a non-pressurized platform [45].

If the thermal control of the satellite at the clock location and the sensitivity of the clocks are known, it is possible to derive the expected effect in the phase obtained from the AFS sensitivity to temperature. In case of agreement between the expected value and the observed harmonics in the 'apparent clock' the source can be conclusively linked to the AFS. In order to review the temperature sensitivity several information is required:

- Peak-to-peak temperature at clock location.

- Sensitivity of the frequency to temperature.

- Amplitude of measured harmonic in orbit and beta-angle correlation.

No temperature telemetry is publicly available, nevertheless some publications contain meaningful information. Each GNSS and clock technology is reviewed hereafter in order to collect this information.

GIOVE-A RAFS

In-orbit temperature of the physical clock is available in the GIOVE mission. It is possible for GIOVE-A to analyze directly whether the harmonics have any correlation with the frequency standards or the payload chain. The temperature for each unit is measured by the platform at the thermal reference point indicated by the manufacturer. RAFS sensitivity to temperature is specified by the manufacturer in fractional frequency as $\pm$5E-14/°C for an expected temperature operation of -10°C to 15°C [154]. By design, the RAFS can operate up to +15°C with a margin up to +20°C for the qualification. Some small margins were introduced to warrant good operation also during qualification. If the maximum temperature is reached, the thermal regulation saturates and the thermal coefficient becomes 10 times higher.

In-orbit a clear correlation between frequency and temperature exists with a value of -1.5E-13/°C in Figure 7.4. This increased thermal sensitivity with respect to the 5.0E-14/°C specifications provided in Table 4.1, is due to the higher temperature outside the designed range with values up to 24°C. The higher operational temperature was expected by the satellite manufacturer and accepted within the objectives of the mission.

Furthermore, the removal of the temperature effects by means of the computed sensitivity and temperature information significantly reduces the observed fluctuations in Figure 7.4(b). This removal allows the clearer identification of a frequency jump at 3E-13 level around DOY 140.

GIOVE-B

The temperature sensitivity of the PHM is slightly lower than for the RAFS with measured values on ground lower than 3E-14/°C [19]. The telemetry sensors trace well the spectrum of changing illumination of the spacecraft during orbital revolution and from solar/lunar eclipses. PHM and RAFS operate well within their nominal temperature range and the temperature at the PHM location is extremely stable (< 0.1°C during one orbit).

Correlation with all payload chain temperatures does not reveal any clear contributor in the satellite payload chain [50]. It can be concluded that temperature variations do not justify the oscillation of 0.5 ns amplitude observed in the estimated phase.

GPS

Limited information is available about GPS clock temperature. Temperature stability at the clock location is estimated to vary with the orbital period approximately in the range of $\pm$5°C for Block IIA and $\pm$2.5°C for IIR/IIR-M [179]. Even higher variations during eclipse periods up to 6°C in 5 hours are also reported by the same author [180].

Typical temperature sensitivities in $\delta f / f /$°C range between 1.2E-13 and 1.2E-14 for ground cesium clocks [18], 1E-13 for space cesium clocks on board Block IIF [181] and 1E-13 reported for Block IIA Rubidium clocks [11].

Based on the temperature sensitivity of the clock and temperature measurements it is possible to remove the temperature effect as demonstrated offline with GIOVE-RAFS in Figure 7.4. This concept has been implemented in the Block IIF TKS with a temperature controller loop to reduce the temperature sensitivity of the RAFS in orbit to 1E-14/°C [11, 138, 181]. The Block

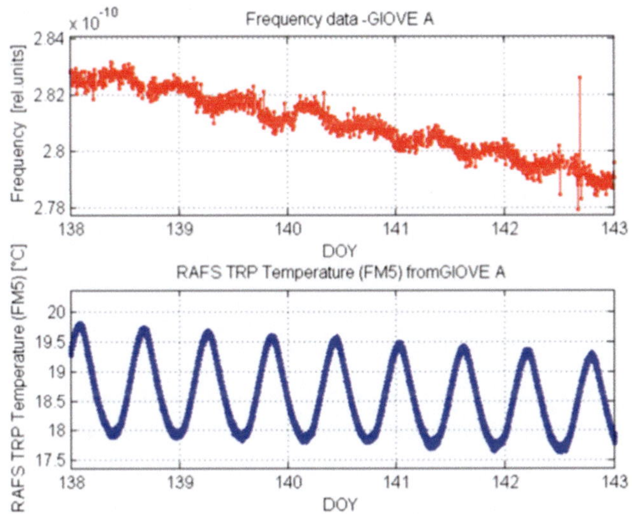

(a) GIOVE-A FM5 'apparent' clock frequency and RAFS TRP temperature, from May 18 to May 22, 2007. Example of periodic fluctuation and a frequency jump observed on the frequency data.

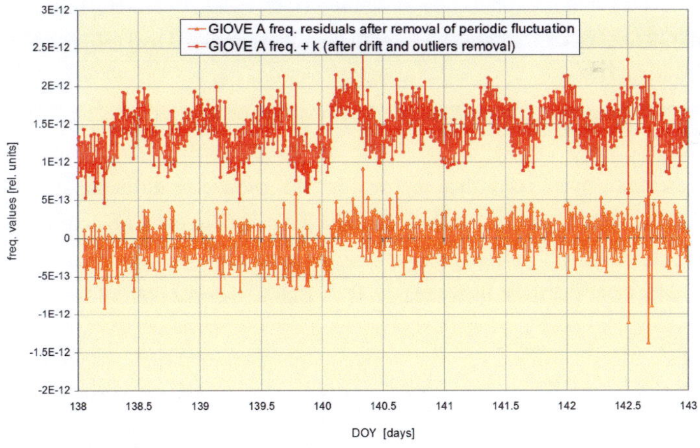

(b) GIOVE-A FM5 'apparent' clock frequency (red dots) and residuals after removal of frequency periodic fluctuation (orange dots)

Fig. 7.4: GIOVE-A RAFS correlation with temperature during eclipse. Source: [65]

IIF Rubidium clock has a temperature coefficient of 2E-13/°C without the benefit of the base temperature controller which improves the temperature coefficient by a factor better than 50. In-orbit temperatures of GPS Block IIF are reported to show thermal variations of less than 0.5 °C peak-to-peak [44]

GLONASS

Few bibliography is available on the satellite temperature at clock location for the different GLONASS block families. Only [13] mentions a variation of $\pm 1°$ C to be expected for GLONASS-M. As a consequence only this block family will be analysed hereafter.

Expected vs measured harmonic

Once the temperature sensitivities S for the AFS and the thermal variation T at the thermal reference point have been collected, it is possible to derive the effect in frequency y and phase (x) of the harmonic; and the expected amplitude be computed from $ST(2\pi f)^{-1}$.

$$\begin{aligned} y(t) &= ST\sin(2\pi ft) \\ x(t) &= \int y(t)dt = ST(2\pi f)^{-1}\cos(2\pi ft) \end{aligned} \qquad [7.4]$$

In Table 7.1 the expected phase oscillation for GNSS satellites from modelled values is compared with the measured in-orbit oscillations observed by POD. Measured GPS values are computed from IGS final clocks, crosschecked with the values reported by [150] and GIOVE values computed from ODTS.

GPS Block-IIA presents a high dispersion with values of up to 8 ns as maximum and other satellites with values as low as 0.3 ns. There are two possibilities to explain this deviation: a different thermal control other than reported for the satellite or different thermal sensitivity for the AFS. The answer can be found in the behaviour of different AFS activated in the same satellite. In case different activated AFS present the same improvement or degraded sensitivity in the apparent clock, the most likely reason is a satellite thermal control better than assumed

Block	AFS	°C	$\delta f/f/°C$	Expected [ns]	Measured [ns]
GPS-IIA	Rb	± 5.0	1E-13	± 3.44	8.0-0.3
GPS-IIR	Rb+TKS	± 2.5	1E-14	± 0.17	0.2-0.1
GPS-IIF	Rb	± 0.3	4E-14	± 0.01	0.5-0.3
GLONASS-M	Cs	± 1.0	2E-13	± 1.73	1.0-3.0
GIOVE-A	Rb	± 2.5	1E-13	± 2.01	2.0-0.8
GIOVE-B	Rb	± 0.5	5E-14	± 0.20	0.5-0.3
GIOVE-B	PHM	± 0.1	3E-14	± 0.02	0.5-0.3

Tab. 7.1: Expected versus observed harmonics

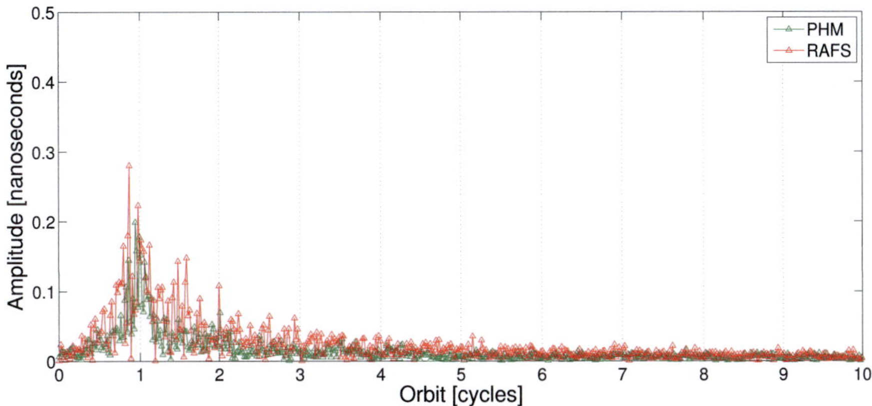

Fig. 7.5: GIOVE-B (E16) FFT with different AFS selected as nominal

(e.g.SVN-17, 29, 31). On the contrary, in case that only one of the apparent clocks presents a different behaviour with respect to the expectation, the probable reason could be a different thermal sensitivity for the particular AFS (e.g.SVN-22).

The presented values for Block-IIR are in overall in agreement with the expectation. However, a major discrepancy is observed for Block IIF where only a 7 ps harmonic amplitude was expected. The expected value has been derived from the thermal sensitivity of the RAFS ($2E-13/^{\circ}C$) improved by a factor of 50 due to the base temperature controller. In case the pure thermal sensitivity of the RAFS is used without this improvement factor a maximum of 0.17 ns would be expected, still below the measured values. As a consequence, either the temperature at the AFS is higher than reported, or the thermal sensitivity is still higher or there is an additional contribution. As variations for the SVN62 GPS satellite carrying first L5 frequency are not expected from pure ground tests, the results reported in [88] could be due to group delay variations.

GIOVE-A also presents a good agreement between modelled and measured values. On the contrary, GIOVE-B does not present any agreement for RAFS or PHM with higher values than expected. While in the case of GPS Block-IIA the use of different AFS provided additional information this is not the case for GIOVE-B. The harmonic seems to have the same amplitude independently of the selected AFS. This point is further confirmed in Figure 7.5. The spectra of the harmonics obtained with RAFS (from 10-Jan-2011 till 10-Feb-2011) and PHM (from 01-Nov-2010 till 05-Dec-2010), using a full month of data around the AFS swap as nominal, shows little difference. From these results it can be concluded that the AFS selection has little influence on the harmonic.

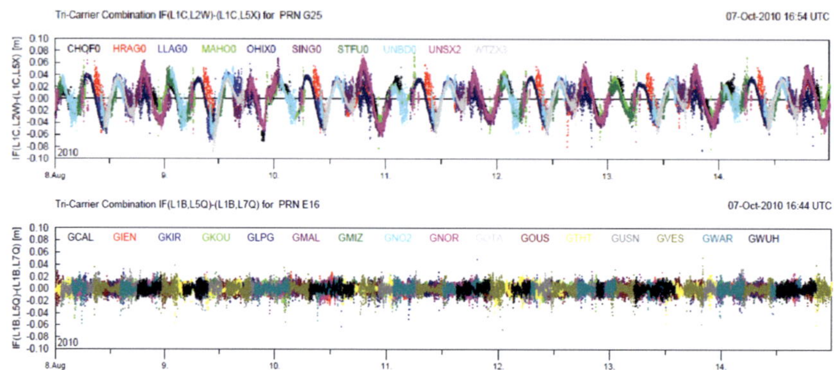

Fig. 7.6: Triple carrier combinations for G25 (SVN-62) and E01 (GIOVE-B). Figure courtesy of DLR (O.Montenbruck).

7.3.3 Group delay variation

The main differences between expected and measured values in Table 7.1 have been observed for the first Block IIF (SVN-62) and GIOVE-B. An additional possible reason for the harmonic source is a group delay contribution. Group delay variations on ground sensor stations are well known specially at UTC time laboratories, the same effect may also exist on the transmitting chain inside the satellite.

For an apparent clock based on double frequency transmissions it is not possible to separate the payload contribution from the pure clock contributions. However, with the arrival of new GNSS satellites triple carrier combinations are possible. Triple carrier transmissions allow, for the first time, the separation of contributions from each signal. The triple carrier combination used by [111] is 'clock free', as dt^s is cancelled, revealing group delay divergences between the signals. In the first real triple carrier transmissions by satellite SVN62 a higher group delay sensitivity to temperature on the L5 signal has been identified as the main contributor to the harmonic in dual frequency combinations. Triple carrier analysis on the first satellite of the Block II-F shows variations in the signal clock mainly attributed to the L5 signal [111], though some contribution should be expected from L1 and L2 signals.

GIOVE-B transmits three separated frequencies E1, E5a and E5b. Figure 7.6, courtesy of the German Aerospace Agency (DLR), presents a clear oscillation for SVN-62 and a flat behaviour for GIOVE-B for the same period. Nevertheless, for GIOVE-B the E5a and E5b signals follow the same path and any fluctuation in the group delay would affect equally both signals, making this combination not suitable to analyze the group delays for GIOVE.

7.4 GIOVE-B special case

7.4.1 Phase meter comparisons

PHM and RAFS are compared through the CMCU unit. CMCU phase meter comparison of the RAFS against the PHM as reference on Figure 7.7 presents the phase meter comparison and Allan deviation between the nominal PHM and the hot redundant RAFS for 25th-27th December 2009, when the fractional frequency offset between both clocks crossed the zero offset. This result does not allow us to draw conclusions on the PHM as the RAFS noise is higher than the PHM and the harmonic effect (1E-13), but it confirms the good thermal control at RAFS location, since a clear signal is observed without harmonic contribution.

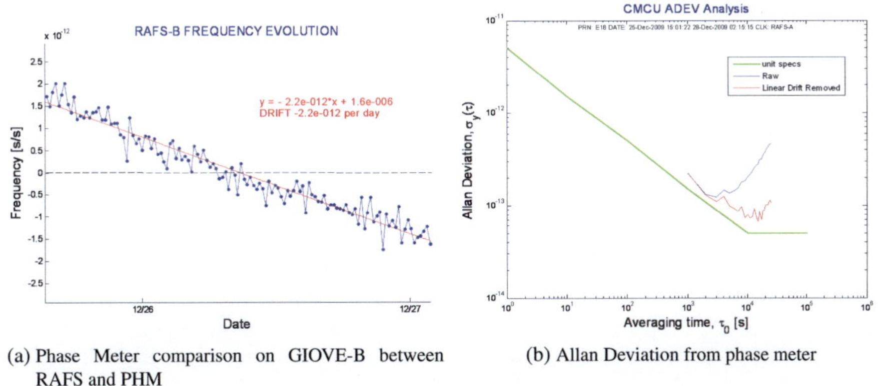

(a) Phase Meter comparison on GIOVE-B between RAFS and PHM

(b) Allan Deviation from phase meter

Fig. 7.7: GIOVE-B on board PHM-RAFS phase meter comparison

7.4.2 Orbit residual

The harmonic observed on GIOVE-B clocks is the only one not explained by the sensitivity of the AFS or by group delay variations. This satellite is further investigated hereafter.

The current harmonic in the apparent clock behaviour observed on GIOVE-B was accurately predicted before the satellite launch by preliminary studies. The harmonic effect was also expected and associated to an orbit residual of 10 cm (or 0.3 ns) linked to a limited number of sensor stations. The predicted Allan deviation in [26, picture 6] is similar to the one observed in orbit in Figure 6.23

In Section 6.3.6 it was explained how the orbit accuracy depends on the number of measurements and therefore on the number of stations. It remains the possibility to attribute some or most of the clock harmonic to the orbit residual. In order to increase the orbit accuracy two options have been studied. First, together with the L-Band measurement, periods with a larger number of SLR measurements have been used in the adjustment with almost no weight (de-weighted) and in a posterior run with weights in accordance with the measurement standard deviation in order to look for any improvement of the results. Second, additional CONGO stations have been added to the estimation and processed with NAPEOS.

For the first approach using SLR data, two different software packages have been applied: GIOVE-M (run by GMV) and BAY-PAF (Bernese run by Astrium). The Figure 7.8 shows the detrended clock phase estimated by Astrium-Germany using BAY-PAF with weighted (black) and de-weighted (red) the SLR measurements. Up to 46 SLR measurements were included in this five days arc for 7-11 December 2008 (doy 342-346), coming from 5 stations. Both software recover the same harmonic for each run with some reduction in the amplitude. A consistent reduction of the fluctuation between 24% and 54% occurs for GMV and Astrium estimates when SLR observations are weighted in the process.

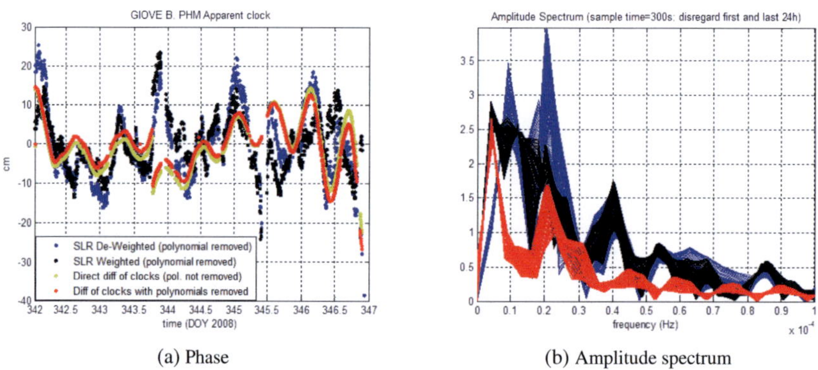

(a) Phase   (b) Amplitude spectrum

Fig. 7.8: GIOVE-B (PHM) phase clock obtained with and without weighting the SLR measurements for 11-15 December 2008 (first and last 12h disregarded). Source: Astrium Germany

The second approach consists of increasing the number of observations by additional stations, since as considered in previous Section 6.3.6 the precision depends on the number of stations. A common period of GESS (13) and CONGO (8) stations have been processed together by ESOC with the NAPEOS software, amounting for a total of 21 stations. The data covers a period of 28 days from mid August 2009 to mid September 2009 ahead the eclipse season for GIOVE-B. Figure 7.9 presents the phase evolution and the spectra of the signal. Two different solutions have been estimated : GESS+CONGO network and only with GESS stations. The clock from each solution plus the difference have been detrended per day and stacked. Two harmonics are recovered by both solutions with a main component of 0.43 ns amplitude on the orbit period (14.0865 hours) and a secondary component of 0.11 ns on the half orbit period (7.04324 hours). The clearer spectrum is obtained with all the stations while some spectral

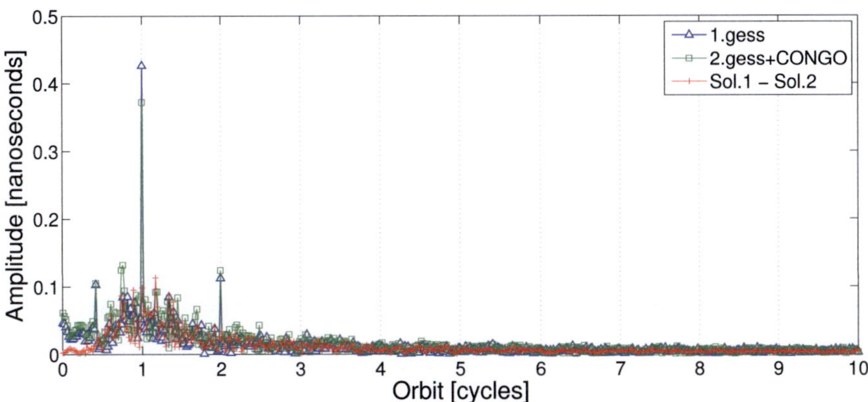

Fig. 7.9: GIOVE-B (PHM) FFT with GONGO and GESS Networks

leakage is obtained with only the GESS solution. The components have the same period with slightly different amplitude between the solutions. The second solution with the higher number of stations reduce the harmonic amplitude and also the residuals of the clocks overlapping, although not as much as expected from Section 6.3.6 study with IGS stations.

The difference between the two solutions with both methodologies shows a clear pattern and fluctuation, even if each apparent clock is noisier. Even if the two harmonic components cancel out in the difference a large amount of energy remains with components at the frequency around one orbit cycle. Nevertheless, both methodologies show an improvement by including more measurements in the estimation.

An additional indication of the limited accuracy of the orbit due to the limited number of measurements derives from the arc length used in the estimation. While 1 day arcs is the typical duration in IGS processing, up to five days length have been identified as the most suitable length in GIOVE mission from the quality of the RMS difference between one day overlapping arcs. The same arc length has later been used by ESOC and DLR for GIOVE estimations. In case 5 days arcs estimation is used to improve the orbit, some 'butterfly' or 'bath-tub' effects are observed at the borders with larger differences between the weighted and deweighted solutions and it becomes necessary to extract only the 1 day central arc.

Some interesting feature of the harmonic is the slight difference (+2 m 50 s) between the harmonic (14.0865 h) and orbit period (14.03916 h $\pm$ 5 s) estimated by:

$$\frac{T^2}{a^3} = \frac{4\pi}{GM_E} \qquad [7.5]$$

with the mean semi-major axis transmitted in the navigation message during the analyzed time

($a = 29545305.8$). Beside further discarding the temperature as root cause of the harmonic, the difference could provide some indication about its origin. Similar differences (+1 min) have also been reported for GPS satellites [150], however this is a particular interesting analysis to be performed with PHM clocks due to the less noise signal and the low probability of temperature effects in the apparent clock.

In summary, any data addition by arc length increase, SLR data or sensor station, improves the orbit quality and decreases the harmonic amplitude. This fact indicates that some orbit residual still exists in the apparent clock affecting the harmonics amplitude. Additionally, a difference (+2 m 50 s) between the harmonic and orbit period exist which could help to identify its origin.

7.4.3 Argument of latitude dependency on SRP

After the harmonic origin has been finally attributed to the orbit accuracy, now the final question is what in the orbit estimation generates this orbit period dependency. One possibility could be the SRP model coefficients estimated together with the orbit.

The instantaneous Solar Radiation Pressure (SRP) is an inertial acceleration component due to direct solar radiation pressure upon the satellite. This force $\vec{a}_{SRP}$ needs to be included in the equations of motion of the satellite as explained in Section 5.3.2. In GIOVE mission the POD is performed using an empirical SRP model (Equation 7.6 presented in [136]) based on [23] with additional harmonics in all three directions with a total of 9 parameters ,

$$
\begin{aligned}
\vec{a}_{SRP} = \quad &D_0\vec{e}_D + D_c\vec{e}_D cos(u) + D_s\vec{e}_D sin(u) + \\
&Y_0\vec{e}_Y + Y_c\vec{e}_Y cos(u) + Y_s\vec{e}_Y sin(u) + \\
&B_0\vec{e}_B + B_c\vec{e}_B cos(u) + B_s\vec{e}_B sin(u)
\end{aligned}
\qquad [7.6]
$$

where $\vec{e}_D$ is the unit vector satellite-Sun, positive towards to Sun, $\vec{e}_Y$ is the unit vector along the spacecraft's solar-panel, positive following the definition of the satellite reference frame, and $\vec{e}_B$ is the unit vector which completes the right handed system. Figure 7.10 shows the reference frame used for the empirical SRP model. The empirical parameters of the model to be estimated are D_0, D_c, D_s, Y_0, Y_c, Y_s, B_0, B_c and B_s. An additional second harmonic model, where 15 parameters are estimated, were tested in GIOVE mission with good results. Nonetheless, a generally observed improvement was not always consistent between the arcs and it was recommended to be reviewed when more stations will be available [50].

The main acceleration $D_0\vec{e}_D$, along the Sun-Satellite direction, represents most of the SRP acceleration, since the rest of the terms take values of around 1% or less of the magnitude of the main component. The model contains two harmonic functions of the argument of latitude u along the three directions.

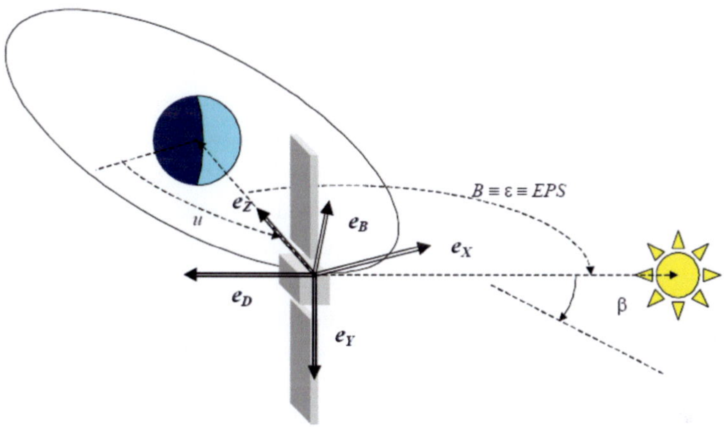

Fig. 7.10: Reference frame for the empirical SRP model. Source: [136]

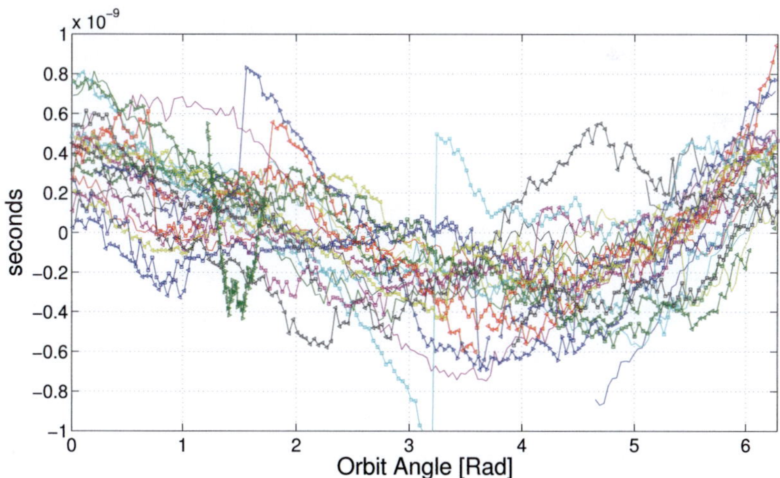

Fig. 7.11: Harmonic as function of argument of latitude angle

The apparent clock harmonic is correlated with the argument of latitude angle as depicted in Figure 7.11. The figure has been obtained after a day by day by detrending the phase data and plotting each orbit versus the argument of latitude. Additionally, the independent clock estimations by the GGSP consortium did not agree only for the PHM (previous Figure 7.1(e)). The SRP model used by each analysis center for each estimation was slightly different what could justify also the disagreement observed in the clock. Both facts indicate an inaccuracy of the SRP model as the most likely cause of the harmonic observed in GIOVE-B PHM clock.

Nevertheless, it must be highlighted that SRP modelling suffers from an observability problem due to the reduced coverage (along and across components are difficult to observe) and therefore, independently of the model being used, the data quality is the main driver for a correct satellite dynamic predictability. This hypothesis will be confirmed for Galileo IOV satellites once enough IGS stations become available to compute the orbit with sufficient accuracy. As an indication, IGS analysis centers use in average around 100 stations for the estimation of GPS satellite orbits.

The period for Figure 7.11 has been selected during eclipse season when the SRP estimation is less accurate due to the lower beta angle, as acknowledged in [12]. Each epoch used for the clock phase comes from an average of 48 different estimations using a moving arc estimation of 48 hours with one hour step between the arcs. All other periods outside eclipse present a clear correlation with the argument of latitude. The effect of temperature being excluded, it has been selected to additionally demonstrate how the PHM allows the identification of orbit modelling errors.

7.5 Conclusions

The only unknown effect in the PHM behaviour is a 0.5 ns harmonic observed in the apparent clock. While harmonics in GPS satellites have been a well-known feature since their early estimation [158], with only one recent publication [150] mentioning the temperature as the origin of this effect, lacking any dedicated analysis. This effect was required to be understood in order to apply corrective measures to the clock design or to the estimation if confirmed to be an artefact of the processing.

The origin of the harmonic in the apparent clock of GNSS satellites has been reviewed and clarified in this chapter. First, it has been demonstrated how the amplitude is correlated with the sun-beta angle for the majority of the satellites. This correlation indicates a possible dependency on temperature. Second, a methodology has been proposed in this thesis to derive the expected amplitude of the phase oscillation due to the sensitivity of the AFS. Public information has been collected for all GNSS to apply this methodology and to compute a-priori values. The expected values have been compared with the measured values from POD. It has been proved that the measured amplitude for almost all GNSS satellites is in good agreement with the expected values. This good agreement indicates that harmonics in the apparent clock of GNSS satellites are mainly due to the thermal sensitivity of the AFS.

Disagreement has only been observed for GIOVE-B (PHM and RAFS) and SVN62 (Rubidium Block IIF). In the case of GPS satellite SVN62, the harmonic source has been demonstrated to be due to the group delay; however, for GIOVE-B, temperature-induced variations in the AFS seem unlikely in view of numerous pieces of evidence: the missing correlation between on-board temperature and frequency, the small amplitude of temperature variations on the PHM, the low dependency on sun-beta-angle, the same spectra observed when RAFS is selected, and the poor agreement between independent estimations.

It seems that if other error sources cause the orbit-periodic variations in the clock data for GIOVE-B (PHM). Degraded orbit accuracy seems the most likely source of the variations, as it was tentatively predicted before the satellite launch for the limited envisaged amount of 13 stations. This hypothesis is demonstrated by the reduced amplitude (25%) when increasing the number of stations (+8), including SLR measurements or by the better (rms) repeatability when extending the arc length till 5 days.

The SRP coefficients are dependent on the argument of latitude and, as a consequence, also on the orbit period. The SRP seems to be the probable cause of the harmonic in the PHM estimation. The empirical model used for SRP estimation may be inaccurate, or the coefficients estimation is affected by the degraded geometry due to the low number of stations. Once the Galileo constellation is deployed, the accuracy of the SRP for Galileo satellites in PHM mode should be reviewed when a higher number of sensor stations becomes available.

8 GNSS clock stability and prediction

8.1 Introduction

Satellite AFS are heavy, power-hungry consumption and expensive pieces of hardware, especially when compared to their equivalent terrestrial crystal oscillators used in the majority of receivers including many reference stations and their space qualified Ultra Stable Oscillator (USO) versions used in many satellites. Crystal oscillators (XO) in the passive AFS used in GNSS provide the performance at short term ($<$1-10 second). Since temperature influences the operating frequency, various forms of automatic compensation are employed in the design, from analogue compensation (TCXO) and microcontroller compensation (MCXO) to stabilization of the temperature with a crystal oven (OCXO). Analogously satellite AFS require thermally controlled base plates and radiators to stabilize the clock temperature. If commercial oscillators are already used in other satellites, why launch dedicated atomic standards into GNSS satellites increasing the cost and complexity of the satellite design?.

Three characteristics of AFS provided in Section 4.2 can be identified as an answer to this question. First, their predictability associated to a better long term stability makes it possible to accurately predict the clock model provided to the user for real time navigation. Second, their reliability means fewer operations. And third, their several orders lower drift requires less timekeeping maintenance and fewer interventions from ground - for example, Spectratime USO has a fractional frequency drift of 3E-8/year, RAFS 3E-10/year and PHM 3E-12/year.

Clock offset $dt^s(t)$ prediction at the time of user PVT still represents one of the major error contributors for real time navigation. Clock corrections are also the main added value of services based on real time double frequency measurements (as IGS-RT or FUGRO).

In this chapter, the overall GNSS clock prediction is analyzed. First, the performance is reviewed in terms of stability. Then, events affecting the clock prediction robustness such as frequency steps or clock maintenance are identified and analyzed. These events especially disturb any integrity applications (e.g. civil aviation). Finally, current clock prediction strategies and relevant integrity methods are reviewed and applied to the satellite 'apparent clock'.

8.2 Clock stability

The Allan Deviation obtained for a given time interval is the classical figure of merit to report clock stability. Figure 8.1 shows the Allan deviation for all transmitting GNSS 'ionosphere-free

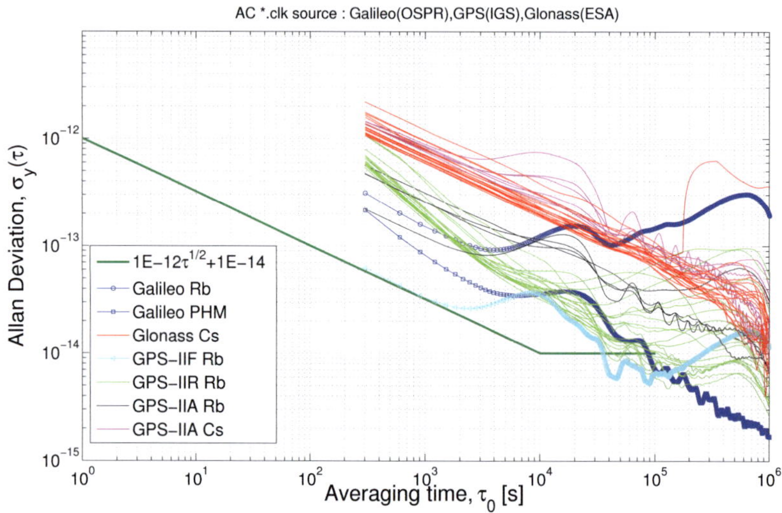

Fig. 8.1: Allan Deviation for GNSS clocks during October 2010

clocks' during October 2010. Special attention needs to be placed on the data source used to compute the frequency stability. Final IGS 'ionosphere-free clocks' solutions are used for GPS, ESA (IGS-AC) for GLONASS and GIOVE mission for Galileo.

Some limitations should be taken into account before extracting conclusions. IGS estimations are obtained for around 30 satellites with almost 200 ground stations. Galileo clocks are estimated with only 2 satellites and 13 stations with experimental receivers. As a consequence, the GIOVE 'ionosphere free clock' estimation noise is higher than for IGS. The timing signal noise is lower than demonstrated in previous sections. GLONASS clocks suffer from boundary jumps linked to the different receiver hardware delays associated with the FDMA signals - their stability being better than reported. Nonetheless, the figure presents the current state of the art of GNSS time transfer with the current limitations to estimate satellite time scales based on FDMA signals.

Clocks in Figure 8.1 may be ranked in families from the least to the most accurate. Cesium clocks on GLONASS and GPS have the poorest stability. GIOVE free running rubidium clocks have better short term stability than GPS-IIA rubidium but the instability of the drift rate together with the thermal sensitivity on GIOVE-A provides a lower level of performance than their GPS equivalent. In Block-IIF, the signal clock derived from the TKS from the rubidium has a higher short time noise but different for each specific unit in the long term. Finally, PHM and GPS-IIF rubidium clocks provide the best performance. The new frequency standards flown in GIOVE satellites need a more thorough review in order to extract conclusions for the coming Galileo system. Consequently, a detailed review of GIOVE clocks is here performed.

8.2.1 GIOVE RAFS

A total of one engineering qualification model (EQM) and six RAFS flight models (FM) were manufactured for the GIOVE mission. Four of these clocks were mounted on-board the GIOVE-A and -B satellites. The EQM and FM2 clocks are still maintained on ground at ESTEC(NL) for testing in a mock-up of the satellite payload. FM3 clock is kept as a spare at the manufacturer's premise. Table 8.1 collects the serial number (SN) of the clocks manufactured for the GIOVE program and their allocation to the satellites.

SN	Model	Location	TM
001	EQM	ESTEC	-
002	PFM	GIOVE-B	RAFS-B
003	FM1	GIOVE-B	RAFS-A
004	FM2	ESTEC	-
005	FM3	SpT	-
006	FM4	GIOVE-A	RAFS-A
007	FM5	GIOVE-A	RAFS-B

Tab. 8.1: GIOVE-RAFS-list

The two RAFS on board GIOVE-A operated outside their expected temperature range as clarified in the previous chapter. As a consequence, GIOVE-A clocks should not be considered fully representative of the family due to their operation outside the specified operational temperature. Nonetheless, as no anomaly has yet been detected on the clock through telemetry or via a signal, the functionalities being as expected, FM4 and FM5 units on board GIOVE-A will be considered as being inside the family bearing this limitation in mind.

Figure 8.2 shows the GIOVE-A clock behaviour for the period 2007-2011. The first subplot (a) presents the fractional frequency for the complete period. Each colour represents a continuous operation period of the clock. The second subplot (b) shows the fractional frequency drift rate by period from switch-on-time. It is observed that between 5 and 30 days are required to stabilize the drift below 1.E-12/day depending on the period and clock. Finally the third subplot (c) provides the dynamic Allan deviation at different integration times. The clock and signal do not operate in a continuous mode, the figure presents spikes caused by discontinuities in the time scale due to changes in the nominal clock selection, signal transmission mode or ground station reference. Frequency steps also affect the stability at longer intervals. Nevertheless, the noise level is stationary without showing any sign of degradation after 5 years of intermittent operation in orbit.

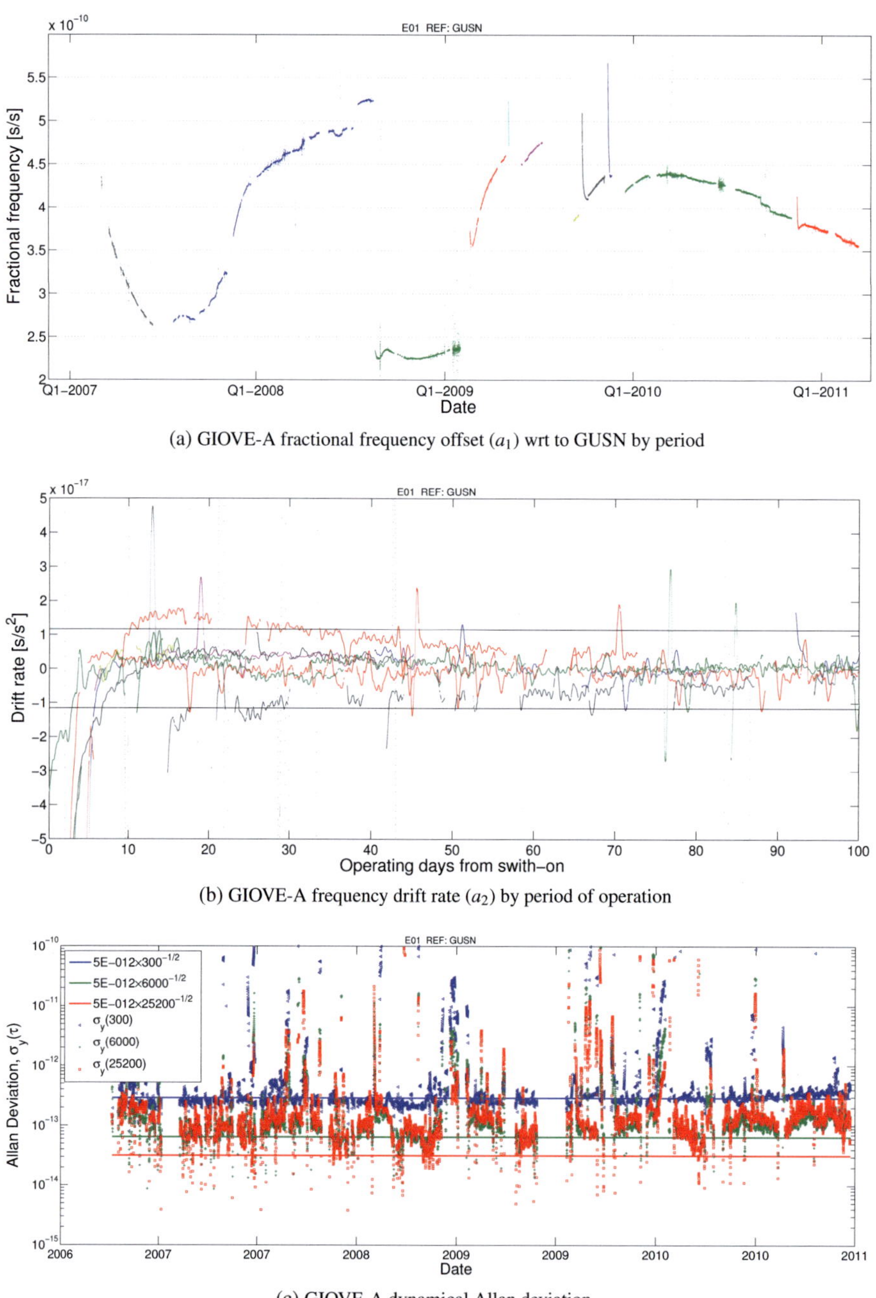

(a) GIOVE-A fractional frequency offset (a_1) wrt to GUSN by period

(b) GIOVE-A frequency drift rate (a_2) by period of operation

(c) GIOVE-A dynamical Allan deviation

Fig. 8.2: GIOVE-A(E01) clock frequency offset, drift and Allan deviation

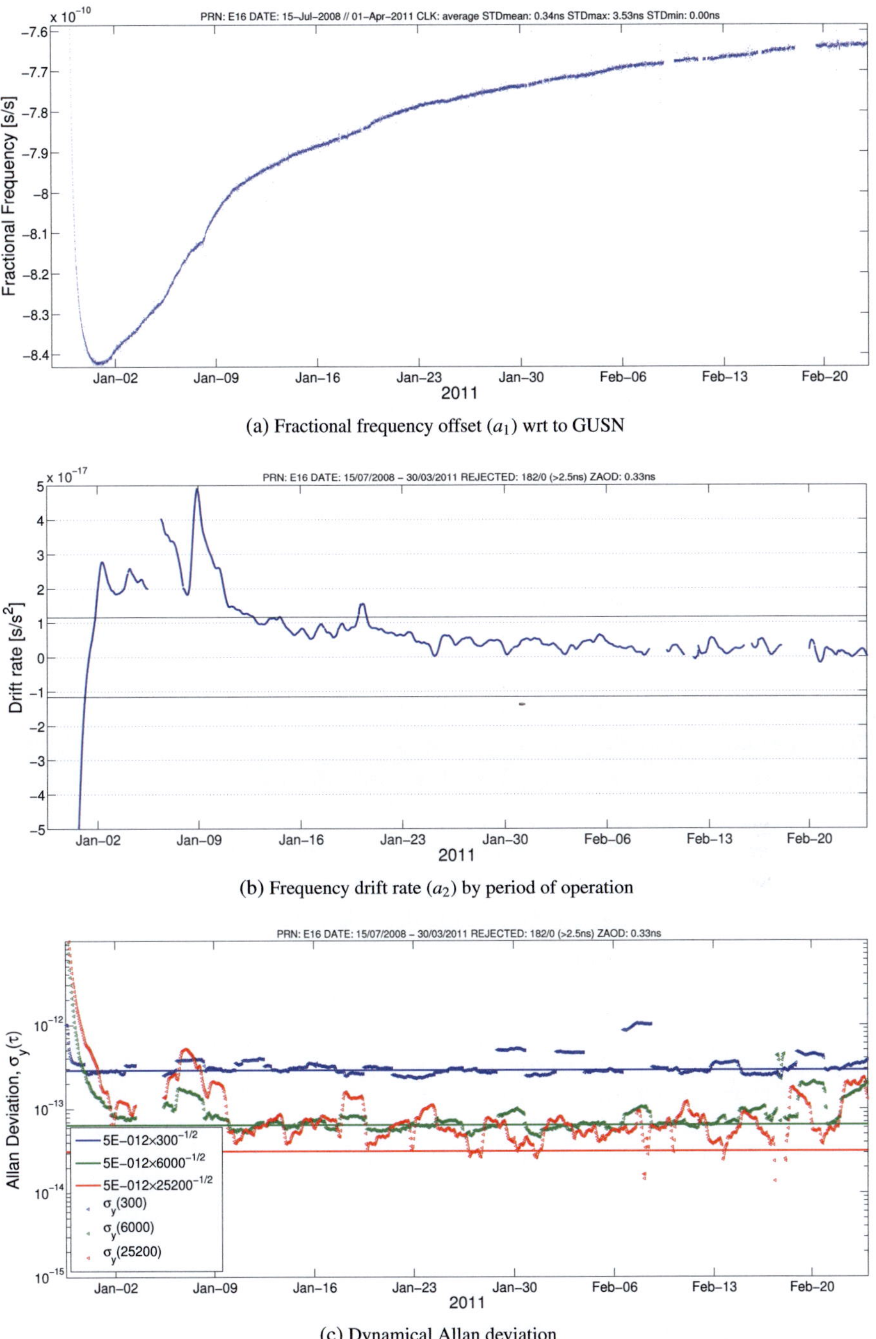

(a) Fractional frequency offset (a_1) wrt to GUSN

(b) Frequency drift rate (a_2) by period of operation

(c) Dynamical Allan deviation

Fig. 8.3: GIOVE-B(E16) RAFS SN002 operation as nominal clock

On Galileo satellites, contrary to other GNSS, it is possible to operate two clocks in parallel from which only one (typically the PHM) is used for navigation. In case of failure of the nominal clock the backup (typically the RAFS) will be ready for use without waiting for any stabilization time. Additionally, this choice allows to gain in flight experience of these two new technologies. GIOVE-B carries on-board a PHM (PFM) and two RAFS units. One clock is selected as nominal for transmission while the other is kept powered on in a hot redundant configuration. The PHM was the nominal clock for transmission since the launch. After 2 years of GIOVE-B operation on PHM mode, the RAFS-SN02 was switched on on 29/12/2010 and selected as nominal clock for transmission. The fractional frequency offset, drift rate and stability have been shown in Figure 8.3. This unit presents the typical stabilization process of the family following the switch-on after a long non-operational period. Immediately after the clock locks, the fractional frequency (a_1) retraces to 1E-10 and the Allan deviation stabilizes after two days to the ground measurements level $5\text{E-}12\times\tau^{-1/2}$. The frequency drift rate (a_2) stabilizes below 1.E-12/day after 16 days of operation.

8.2.2 GIOVE PHM

The PHM-PFM was the nominal clock on board GIOVE-B during most of the operational time. Table 8.2 covers all operational periods for PHM with on/off dates, duration, frequency drift rate and the frequency retrace after each switch on.

The frequency retrace is the difference between the observed fractional frequency values before and after a switch-off cycle. The clock does not drift during the time that it is off which indicates that the aging is due to some effect in the physical package which happens only during its active operation. A frequency retrace is computed by fitting a linear drift to each period, and by simply comparing the end of one period with the beginning of the next using the telemetry on/off times. The frequency retraces between two subsequent on-cycles is typically a few parts in 1E-12, improving over time until the lower 1E-15 value is observed in the last (5th) operation period. This property was already identified during ground tests and allows the verification of the relativistic frequency shift at the 1.2 % percent level in Section 3.7.5.

#	On	Off	days	a_2 (df/f/day)	df/f retrace
1	05/05/2008 17:47	03/06/2008 16:02	28.9	-7.20E-15	
2	09/06/2008 21:09	25/06/2008 07:43	15.4	-3.69E-15	-1.02E-13
3	05/07/2008 15:02	08/09/2008 15:51	65.0	-2.93E-15	1.52E-14
4	24/09/2008 08:50	04/11/2009 11:40	406.1	-8.73E-16	8.65E-14
5	11/11/2009 08:53	07/12/2010 13:50	391.2	-6.87E-16	-4.59E-15

Tab. 8.2: PHM operation periods

The PHM in orbit appeared to have an extremely low frequency drift and its long-term performance is analysed when being referenced to a steered active hydrogen maser. The GUSN

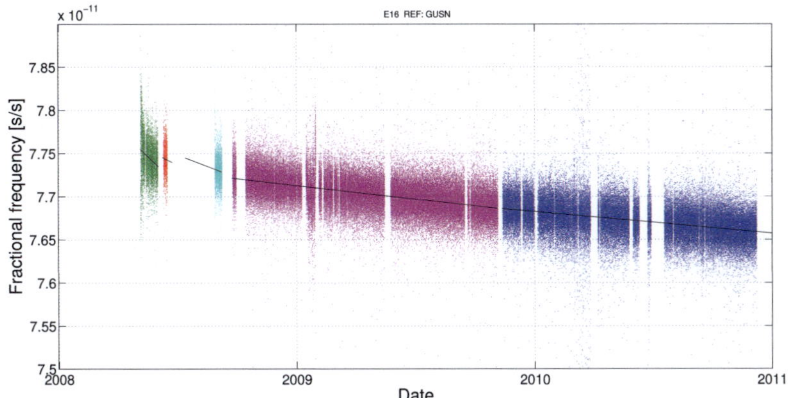

Fig. 8.4: E16 (PHM mode) Fractional frequency (a_1) by operating period referred to GUSN

station is selected as reference to refer to all clock estimations. GUSN uses the same reference frequency source as USN3 which is connected to the reference signal from the USNO Master clock (MC2), the primary realization of UTC(USNO). Figure 8.4 shows the fractional frequency offset of the PHM on-board GIOVE-B as estimated by the ODTS from the first switch-on until the end of October 2010. The various colours correspond to the 5 continuous periods of PHM operation described in Table 8.2.

The overall general trend of the PHM is quite different from the one of the RAFS. First, its fractional frequency variation just after switch-on is extremely stable and does not show any sign of non-linear equilibration processes. This is believed to be due to the intrinsic PHM technology that is less sensitive to long-term physical equilibration processes. Second, traceability is significantly below the 1E-10 values observed for RAFS.

The PHM performance has also been analysed in terms of Allan Deviation, based on two different methods. The first one relies on the direct processing of the ODTS estimated phase offsets. As the ODTS provides a clock estimate every 5min (300sec), the Allan deviation cannot be estimated for integration times below this value. The second method is based on the direct processing of the One-Way Carrier Phase measurement (OWCP), and is detailed in section 6.5. Figures 8.1 and 6.23 present the Allan Deviation of the PHM on-board GIOVE-B as estimated by these various methods.

These figures show that, as anticipated, the short-term stability of the PHM estimated by the ODTS is limited by the system noise at short-term. Beyond 3000 seconds, the estimation is affected by a periodic oscillation at the orbital period. This effect is analyzed in chapter 7.5 dedicated to clock harmonics, being considered to be mainly due to a limitation in the orbital models with a limited number of sensor stations. It can be concluded that PHM frequency stability and long term drift are the best among GNSS satellites, with an excellent frequency traceability up to 1E-15 (Table 8.2) and shows the lowest thermal sensitivity (Table 7.1). These characteristics allow the identification of the PHM as the best operating GNSS clock in terms of performance, providing new possibilities for navigation.

8.3 Clock events

Main clock events affecting the prediction are associated to failures, frequency steps and maintenance operations. Failures are, by their very nature, unexpected; nevertheless, one of the advantages of AFS is the possibility of detecting failures in advance from the telemetry or progressive degradation of the timing signal leading to maintenance operations in order to change a degraded unit. Frequency steps and maintenances are analyzed hereafter in order to derive the potential effect in the prediction strategy.

8.3.1 Frequency steps

It is commonly known that some rubidium clocks generate frequency steps, and the frequency steps for GPS-RAFS tend to decrease in size and rate of occurrence over time [47]. These jumps were referred to as pre-ageing behaviour (mechanical relaxations). According to the literature the drift stabilisation period of a RAFS may last 100 days. A different behaviour in terms of step magnitude and occurrence is observed between units of the same family which could indicate some link with the final realization of the physical package. Frequency jumps in GPS satellites are reported for one third of the RAFS, being typically below 1E-12 with a yearly frequency. Step size, shape and occurrence in GPS rubidium clocks depend on each single unit. The same characteristics seem also to be true for Galileo clocks.

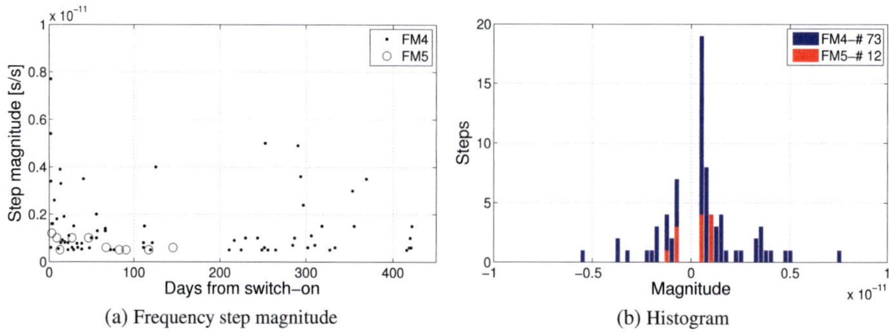

(a) Frequency step magnitude   (b) Histogram

Fig. 8.5: Frequency Steps in GIOVE-A

Figure 8.5 summarizes all frequency jumps observed on GIOVE-A for each switch-on cycle. Jump amplitude, start and duration were extracted manually by visual inspection. No automatic detection functionality was used. A threshold of 5E-13 was applied to detect the jumps which sometimes seems rather arbitrary, leaving without reporting numerous jumps on this order. In general, it can be concluded that manual extraction makes it difficult to distinguish between the coupling of small jumps below 5E-13 and the frequency oscillation associated to the temperature sensitivity on RAFS.

Steps are more frequent during the first 100 days of the stabilization period, as can be seen in Figure 8.5. FM4 seems to be out of family with magnitudes up to 1E-11, an average of 10 days between steps and no decrease of magnitude occurrence in time. FM5 presents an overall better behaviour - once the drift stabilization period has been achieved, the frequency jump magnitude decreases to values below 5E-13 and one per month occurrence. Unit tests on ground have shown similar or better behaviour of FM5 in size and occurrence. One third of the units have not presented any step during the limited time of the ground tests.

The clock model provided to the user in the navigation message is defined by a second order polynomial (Equation 3.4). Assuming a constant initial frequency offset, drift rate and no noise, the impact of the frequency step in the UERE can be modelled through a drift rate step (a_2) of magnitude δa_2 over a time span from t_0 (step-starts) to t_n (step-ends). The concept is illustrated in Figure 8.6. Real steps in the drift rate can be observed in the drift rate estimations for GIOVE clocks in Figures 8.2 and 8.3.

The integration into the a_0 term can be assumed, in a worst case approach, as the error after the end of the step at t_n till the next update of the navigation message at t_{oc} for a user at a given time t.

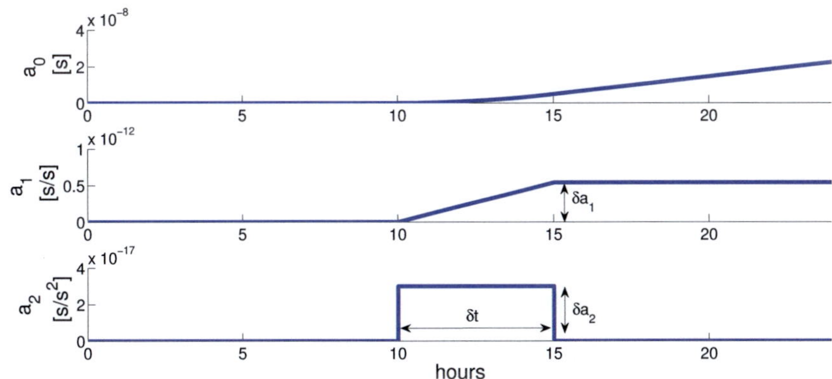

Fig. 8.6: Frequency step components

$$UERE(t_{oc}) \quad = \delta a_2(\delta t)(t_{oc} - t_n) + \tfrac{1}{2}\delta a_2(\delta t)^2 \qquad\qquad [8.1]$$

The main drivers in Equation 8.1 are the magnitude (δa_2), the time span of the step ($\delta t = t_n - t_0$) and mainly the time to update the message after the end of the jump ($t_{oc} - t_n$). The step type may be classified as instantaneous ($\delta t \approx 1$ second) or over a time span of typically hours. Instantaneous steps generate a larger error but are easier to detect as spikes in the frequency. Steps over a time span take more time to integrate into a larger error but are more difficult to be detected.

	unit	Galileo	GPS
Frequency of occurrence	[days]	30	365
Magnitude (δa_1)	[df/f]	1E-12	5E-13
Duration (δt)	[hours]	2.5	0
Message update (δt_{oc})	[minutes]	100	720
max URE	[meters]	1.2	6.5

Tab. 8.3: RAFS frequency step impact in the UERE

The maximum UERE impact based on the identified model and typical step is provided in Table 8.3 for both GNSS systems using RAFS. Frequency steps in GPS and Galileo rubidium clocks have different characteristics. GPS steps are rather instantaneous, with a magnitude below 5E-13 and yearly frequency. Galileo rubidium clocks suffer similar steps with longer integration time of up to several hours, larger magnitude 1E-12 and a low rate of occurrence. However, the impact to the user is up to 6.5 meters at the end of the navigation message validity interval for GPS and 1.2 meters for Galileo due to the foreseen shorter update period of the navigation message. These infrequent errors are not visible in the standard performance metrics due to their low occurrence rate, but instead can be seen in the tails of the error distribution function, as will be demonstrated later in Section 8.4.

8.3.2 Clock maintenances

Clock maintenances disturbing the satellite time scale continuity are required due to several reasons:

1. Clock operational maintenance

2. Clock malfunction recovery

3. Satellite maintenances

4. Timekeeping

Clock operational maintenance

Clock operational maintenances are performed for Block-IIA cesium clocks. Maintenances are performed approximately twice per year by pumping of the beam tube to maintain working order. This maintenance requires, on average, 18 hours of unusable time for each satellite as stated in gpsb2.txt

Clock malfunction recovery

Clock malfunction requires a clock switch to replace a misbehaving clock with another redundant unit. Unavailable operational time depends on the ground control segment reaction time to detect the event, remove the satellite of the constellation, correct and then reintroduce the satellite.

Maintenance activities are announced in advance as far as possible by the control segments by Notice Advisory to NAVSTAR Users (NANU), Notice Advisory to GLONASS Users (NAGU) and Notice Advisory to Galileo Users (NAGU). The information distributed in these notices is collected by several users. For example, the gpsbt2.txt file maintained by USNO collects detailed information about the satellite maintenances for GPS, including clock type change. However, some changes between frequency standards or some special maintenances have not been logged, as demonstrated hereafter by analyzing some practical examples.

During the analyzed period, two interesting changes were observed in Block IIA as extracted in Figure 8.7. The first change was in satellite SVN40, which changed from cesium to rubidium at the end of 2007. The rubidium standard did not stabilize during January, suffering a sudden increase in the drift in February-March. The clock was finally replaced in April by a cesium as reported in gpsbt2.txt; the second clock change happened in SVN38, the active cesium being replaced on 16 October 2009 by a rubidium after the frequency instability observed for the cesium during the previous months became more severe. The satellite was then removed from the constellation (nanu.2009083.txt). After the clock swap, the rubidium clock started its stabilization period during which one large frequency step was observed, most likely a commanded adjustment. After sudden degradation of the rubidium frequency stability, the unit was again changed to another cesium. The satellite was not declared operational till mid-December (nanu.2009125.txt). This maintenance in SVN38 is not reported in gpsbt2.txt in which the satellite transmission is considered to be stable during 2009 operating with a cesium clock.

While Block IIA operates one free running clock of the 2 cesiums and 2 cubidiums onboard, the Block IIR uses one of the three rubidium AFS available on-board controlled by the Time Keeping System (TKS). This strategy seems to lead to fewer operations, since only one operational change was visible during 2008-11 period on SVN61/PRN02 in Figure 8.8. This change of clock is interesting due to different steps: it was not scheduled in advance; the notice was sent after 30 minutes of signal interruption (nanu2008044.txt) but no degradation is visible

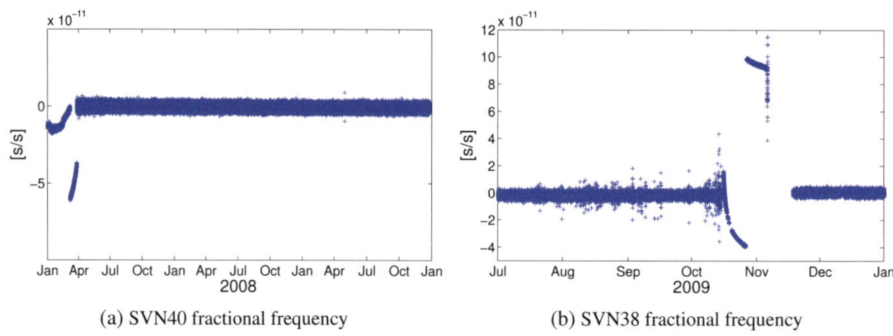

(a) SVN40 fractional frequency   (b) SVN38 fractional frequency

Fig. 8.7: Clock maintenances for SVN-38,-40

in the clock behaviour before the event to justify the deactivation linked to the clock; after switching-off no signal was transmitted between days 26th and 28th; finally, once operational again, the frequency shows the stabilization process typical of a cold start of the rubidium, but the final Allan deviation at 25200 seconds was different. Whatever was the cause for the maintenance, it seems that the control center changed the clock due to some sudden event on-board. It is worth noting that a similar phase offset (a_0) after the event. The RAFS connected to the TKS are cold redundant, thus a new phase offset should be expected as there is no reason to re-synchronize to the last value. It seems that the navigation time scale is not lost in the TKS but maintained by the VCXO when the prime clock is replaced by a cold redundant unit.

Satellite maintenances

Besides the satellite timekeeping also a station keeping manoeuvre may be required to move the satellite back to its original orbital position. In GPS, this is referred to as repositioning or Delta-V manoeuvre. These manoeuvres require, on average, 12 hours of unusable time for each satellite. For Galileo it is expected to have, as a maximum, one satellite repositioning event during the satellite's lifetime.

Time-keeping

Timekeeping maintenances require the steering of clock phase (a_0), frequency (a_1) or drift (a_2) as explained in Section 3.5.4. These have a larger effect on signal availability for Block-IIA with one maintenance per year and around 6 hours of non-operational time for the satellite. Block-IIR is almost free of timekeeping maintenances due to the frequency steering applied by the TKS as explained in Section 3.5.4. For Block-IIF, the adjustments have been performed before declaring the satellite operational, and no information has been available on the operational

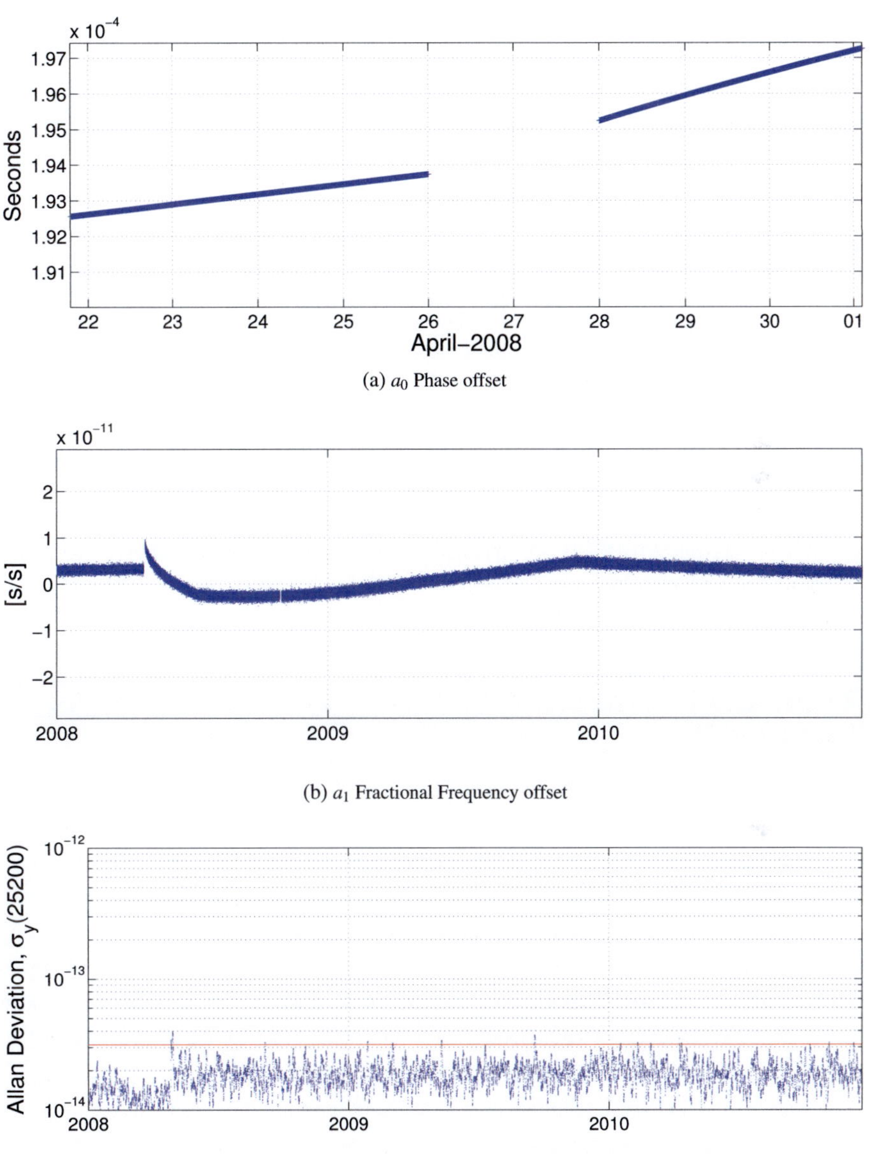

(a) a_0 Phase offset

(b) a_1 Fractional Frequency offset

(c) Dynamical Allan Deviation

Fig. 8.8: GPS satellite SVN61 maintenance in April 2008

strategy and the impact. GLONASS satellites have a limited life time and timekeeping seems to be performed in parallel with other onboard maintenances.

To understand the operational procedure and the impact on the user, the last adjustment for GPS-SVN34 in Figure 3.5 is analyzed in Table 8.4 using IGS final clock products. In this maintenance, the satellite was removed from the constellation for approximately 10 hours and the signal discontinued for 5 hours.

20-Jan-2010 20:16:00	nanu.2010009.txt is issued 6 days in advance to announce the adjustment.
26-Jan-2010 14:00:00	nanu.2010009.txt beginning of maintenance.
26-Jan-2010 15:50:00	Frequency (a_1) is adjusted with the signal transmission on. Frequency offset is changed from -2E-11 to 1E-11.
26-Jan-2010 16:45:00	signal is interrupted.
27-Jan-2010 00:00:00	signal is switched on again. Phase (a_0) has been corrected to zero value with nanosecond accuracy.
27-Jan-2010 00:00:00	nanu.2010009.txt end of maintenance.
27-Jan-2010 00:21:00	nanu.2010012.txt confirmation of end of maintenance at 00:13 UTC.

Tab. 8.4: Analysis of operational maintenance for SVN34

Signal availability

As a consequence of the maintenances, GNSS satellites do not transmit a stable signal for 100% of the time as observed in Figure 8.9. For the three year period from 2008 till end of 2010, the GPS satellites presented a mean availability of the signal in space of 97.33% in Block-IIA and 99.70% for Block-IIR. The availability has been computed based on IGS clock availability for the satellites discarding new launches or decommissioned satellites during this period. In GPS the improved signal availability is mainly due to the change in clock technology from cesium and free running rubidium AFS in Block-IIA to steered rubidium in Block-IIR by reducing the physical and timekeeping maintenances. It is interesting to notice the good availability for Block-IIR for the older satellites, with no major dependency on age.

In comparison GLONASS satellites have a relatively short life time. The expected life time for GLONASS-K and Galileo satellites is 10 and 12 years respectively. However, the availability for current in-orbit GLONASS satellites is lower, especially for models older than 2 years.

GIOVE satellites are an experimental set-up where the signal is changed several times due to test activities. Filtering the expected switch-off periods linked to test activities and taking into

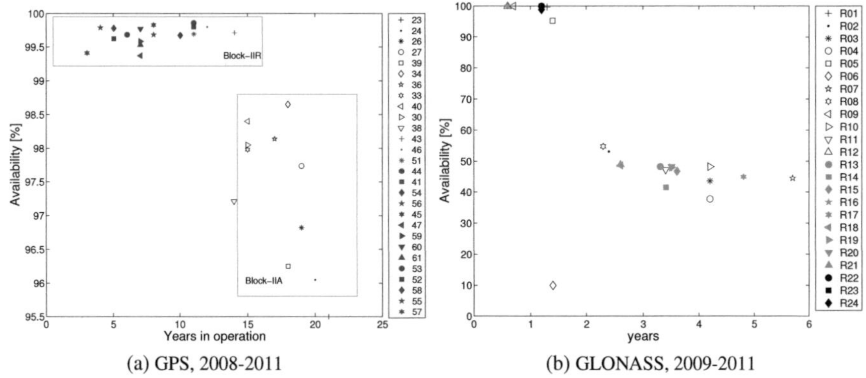

(a) GPS, 2008-2011   (b) GLONASS, 2009-2011

Fig. 8.9: Signal in space (SIS) availability

account the signal in space availability during payload on times, they present an availability of 95.6% for the same period.

In summary, clock and satellite maintenances may have an impact on the clock time scale. Clock prediction strategies for real time systems should take into account partial availability of data, changes in phase and frequency due to timekeeping adjustments or changes of active AFS.

8.4 Prediction

For real time navigation using GNSS, the clock prediction error represents the main error contributor for dual frequency users and the second contributor for single frequency users after the ionosphere, as acknowledged by the error budget of GPS [42]

Since the early GNSS steps, clock error steadily improved as soon as any new clock physical technology become available on board the navigation satellites. From early cesium technologies (GLONASS and GPS), to the first free running rubidium generation (GPS Block-IIA),the second rubidium generation including Time Keeping System technology (GPS Block-IIR),the third improved generation (Galileo and block IIF) until the PHM (Galileo), each technology has brought better clock prediction capabilities resulting in there being a mixture of clock technologies in space.

In parallel to clock performance improvement through better clocks and refreshment rates the prediction robustness gains importance in order to meet International Civil Aviation Organization (ICAO) requirements that would satisfy en route, terminal, and precision approach operations [72]. In this line Satellite Based Augmentation Systems (SBAS) provide complementary information to the broadcast message in order to improve its robustness in terms of accuracy, reliability, continuity and availability. Galileo and GPS evolutions intend to provide their own integrity information. As alternative to SBAS type of integrity information also receiver autonomous integrity monitoring techniques (RAIM) are under assessment to provide a reliable integrity service [168].These techniques analyze the quality of the positioning solution in term of deviations with respect to expected quality. In consequence, information about the accuracy and confidence of the prediction is required to obtain an a priori sigma. Currently just the clock is provided without stochastic information, although previous IGS recommendations following the IGS workshop of 2000 called for the provision of accuracy values for the clock prediction [igsmail-3057] which would allow the user to deal autonomously with different accuracies. This accuracy code was not implemented and will be analyzed here.

The clock prediction strategy starts to get importance in order to improve the accuracy and robustness while dealing with a mixed configuration of clock families or even units, with several efforts in this area. The purpose hereafter will be to provide an overview over the actual GIOVE clock accuracy associated to each clock technology using a common approach and the feasibility to provide accuracy estimation to the prediction. In the following sections the clock prediction strategy and the associated stochastic model will be introduced. Afterwards a reference period will be selected over which different strategies will be applied. A short period of one month is selected to test the different strategies. Once the best strategy is selected for GIOVE clocks the same strategy is applied to all GNSS satellites using IGS data. It is demonstrated how the prediction strategy depends on the refreshment rate; furthermore, the main characteristics are identified and some recommendations provided.

8.4.1 Strategy selection

The clock phase estimations are performed using POD network adjustment techniques by each ground segment to a common reference frame and time scale. To these clock estimations a model is fitted which is uploaded to the satellite and transmitted at a given time of applicability. The user retrieves the message and applies this model to predict the clock offset and compute the navigation solution. The model is transmitted until it is replaced by a new message before the maximum validity time is reached. The concept is graphically depicted in Figure 8.10. In order to achieve an accurate and robust process a dedicated clock prediction strategy is required in terms of:

1. Fitting model

2. Fitting intervals for the model depending on the maximum time of validity.

3. Outlier and rejection of operations over the fitting interval.

4. Overall adequacy to the refreshment rate.

5. Provision of a stochastic model.

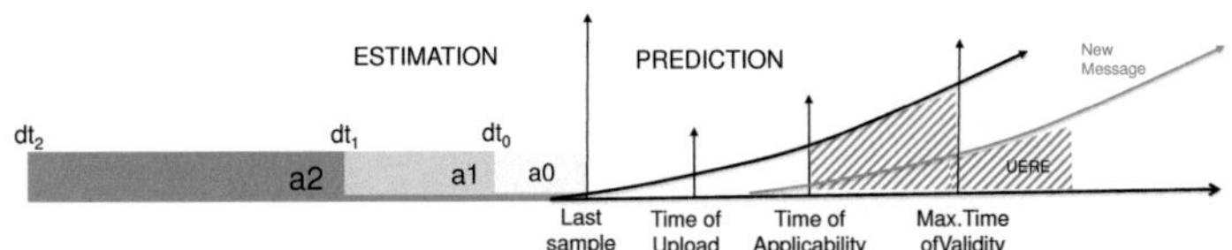

Fig. 8.10: Clock prediction: fitting intervals, prediction error and important time events

First, a linear or quadratic model is fitted to the estimated clocks using least squares adjustment techniques and transmitted to the user using Equation 3.4. The presence of harmonics in the phase restitution of the satellite clock is a well known feature of satellite clocks as highlighted in Section 7.1. A periodic function is recommended to be included by the IGS to the analysis centers in [igsmail-2962], even if it is not clear from the different analysis center reports whether this recommendation is finally applied by each center. The clock model can be extended by using an additional periodic component :

$$\begin{aligned} x(t) &= a_0 + a_1 t + a_2 t^2 + A\sin(2\pi\omega t + \phi) \\ &= a_0 + a_1 t + a_2 t^2 + a_3\sin(2\pi\omega t) + a_4\cos(2\pi\omega t) \end{aligned} \qquad [8.2]$$

where:

- a_0, a_1, a_2, are the polynomial coefficients

- a_3, a_4 , represent the amplitude and initial phase of the harmonic

- ω , frequency, inverse of the orbit period

- t , prediction time from t_0

- t_0 , end of the fitting interval

For each navigation system the model needs to be quantified and included in the allocated space in the broadcast navigation message (as explained in Section 3.5.3). GLONASS, based on cesium clocks, only envisages a linear prediction. The quadratic term, more adequate to describe clock frequency drift associated to rubidium families, is flexibly implemented in GPS and Galileo to be included or rejected depending on the clock drift behaviour of each clock.

The second important parameter in the prediction strategy is the fitting interval to compute the model. A common period is normally used by fitting the clock model to the last 24 hours but a mixed approach can also be used by fitting different estimation intervals for each parameter. For example a_0 can be estimated based on the last hour ($dt = 1$), a_1 on 6 hours ($dt = 6$) and a_2 on the last 24 hours ($dt = 24$) as depicted in Figure 8.10.

Third, any outlier, frequency step or maintenance within the fitting interval will impact the time prediction. Fitting intervals should be pre-processed before final model adjustment. The

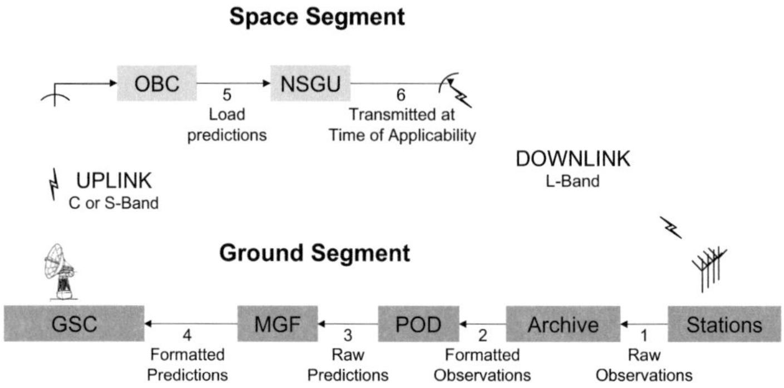

Fig. 8.11: Navigation message generation closed loop

fitting interval is checked before and after the adjustment to remove single outliers. It would also be possible to detect clock maintenances described in Section 8.3.2, adapt the fitting interval in accordance or include additional terms. The outlier rejection strategy deserves dedicated attention and will not be addressed in this dissertation.

The forth and most important parameter after the clock stochastic behaviour is the elapsed time between the observation retrieval and the time of applicability of the new navigation message by the user including the clock model. Clock prediction accuracy is inherently linked to the age of data, that is how old gets the prediction applied by the user with respect to the last data value used in the fitting, the accuracy being inversely proportional to the age. Each GNSS system tries to decrease this age of data from 24 hours in GPS to the 100 minutes envisaged by Galileo or even less by the real time services (e.g. Fugro, Reticle or IGS Real-time service).

The final accuracy for a real time system depends on all the steps required to achieve the full closed loop operation: the latency to transfer the globally collected raw observations in remote locations to the processing center (1), format these raw observations into the archive and retrieve the epoch of these observations by the POD software (2), estimate and predict the navigation message (3), prepare the navigation message for upload (4), upload the navigation message to the satellite signal generation unit (5) and the final time of activation at the next possible navigation message frame(6). The overall closed loop scheme is provided in Figure 8.11. The time to complete this closed loop (1-6) provides the latency rate. Latency should not be confounded with the update rate (3-6), as predictions can be sent every 100 minutes but maybe based on old estimations. Once the navigation message is finally transmitted, its associated error grows with the latency $(t - t_{oc})$ till it is replaced by a newer upload. The user uses only the message between the time of applicability and the maximum time of validity, which corresponds to the shadowed area in Figure 8.10 as also explained by the GPS error budget [42].

The data dissemination rate has improved within the broadcast message. The actual maximum latency according to each system Interface Control Document is smaller than 100 min for Galileo, twice a day uploads for GLONASS and one upload per day for GPS. Upload rates higher than the declared may be employed to adapt the prediction to the clock performance, as in GPS where up to three uploads per day instead of one may be applied in the case of worse performing clocks [42].

Nowadays not only the broadcast navigation message is used for real or near real-time navigation. A global GNSS user may employ full independent orbit and clock information, provided by third entities, such as public precise orbit determination centers as IGS or commercial services as Fugro [107]. These entities provide independent messages by diverse communication channels at different update rates, the almost real time being the current goal [28]. The actual limitation for IGS Ultra Rapid products is not the orbit but the clock prediction accuracy at 9 hours (3 latency plus 6 hours validity) and its robustness with respect to occasional outliers [139]. This limitation is expected to be overcome by IGS Real Time Service.

The update rate is linked to the fitting interval (item 2). Fitting intervals are chosen in accordance to the update rate achievable by the system. Shorter update rates require shorter fitting intervals. Prediction can be avoided for real time applications by using directly the last estimation, provided the clock is estimated at every epoch [67]. Longer update rates require longer intervals, for example, refitting based on several days using previous broadcast messages is applied in some mass market receivers to perform long predictions in order to improve the time to first fix in a warm start [182].

Predictions or estimations can also be avoided. Expert users may compute their own satellites clocks as done in POD adjustments. Analogous to PPP on-line services, this solution is being simplified by internet applications as Magic online service [134] which allow the inclusion of user observations into a global POD solution with minimum interaction of the user.

8.4.2 Experiments

The GPS clock error is still the actual major error source contributor in the error budget. The currently broadcast clock in GIOVE-A and -B navigation messages is computed using a common adjustment to the last 24 hours without any rejection strategy to data fitting. Following this strategy the clock error is also the main contributor in the GIOVE broadcasted navigation message [56]. Different strategies are tested in this section with the GIOVE clock in order to improve the prediction accuracy and the associated stochastic model.

Functional model

Once the model is estimated and the clock predicted, the error x(t) associated with the model can be computed as the difference between the clock prediction and the posterior clock estimation. A standard reference period without events is selected for this analysis. As reference period one month from day 280 to 308 of the year 2009 has been selected. Main attention is given hereafter to the fitting model, the data intervals for the model and the adequacy of the 100 minutes maximum validity time foreseen for the Galileo system. The following strategies are tested and the results are summarized in Table 8.5 :

1. Quadratic fit to the last 24 hours (broadcast strategy in GIOVE-M)

2. Quadratic fit with different fitting intervals (1,6,12 hours) to each coefficient (a_0,a_1,a_2).

3. Quadratic fit with different fitting intervals and two additional components (a_3,a_4) for the periodic component.

    Periodic phase variations associated to clock estimations are a common feature in GNSS. From strategy 2 results, it seems that the harmonic function cannot be neglected without increasing the error. As a consequence, two additional parameters are included in the fitting adjustment (Equation 8.2), where the harmonic period ω^{-1} is fixed to the orbit period.

4. Same as strategy 3, but the periodic terms are not transmitted to the user.

    In order to remain within the 3 parameter model allowed by the broadcast message, the new strategy-4 is tested by fitting the prediction with the 5-parameter model and using only the 3 polynomial parameters (a_0,a_1,a_2) to compute the prediction.

5. Different dt fitting intervals as multiples of the orbit period

    Few information exists about the harmonic source and characteristics. It is not clear whether the inclusion of the two additional terms $(a_3\text{-}a_4)$ in the model will be robust or could introduce outliers increasing the maximum error. Therefore, a simple approach is taken by selecting the fitting intervals (dt1, dt2) multiples of the harmonic/orbit period. In this case, the GIOVE orbit period is around 14 hours and the selected fitting intervals are $dt_1=14$ and $dt_2=28$ hours for (a_1,a_2) respectively.

As summary a final check is performed by evaluating the error at 100 min maximum validity envisaged in the Galileo navigation message with all the strategies under test. Table 8.5 summarizes the results for each prediction strategy, where the first 4 columns represents the strategy applied and the last 4 the associated error observed for PHM and RAFS during the selected period at 100 minutes. For the strategy, the column '#' indicates the strategy number, column

Strategy			RAFS		PHM	
#	dt(0,1,2)	Estimation	BRD	rms	max	rms	max
1	(24,24,24)	a_0-a_2	a_0-a_2	7.31	12.62	0.38	1.46
2	(01,06,12)	a_0-a_2	a_0-a_2	0.97	2.74	0.37	1.43
3	(01,06,12)	a_0-a_4	a_0-a_4	0.70	2.13	0.33	1.73
4	(01,06,12)	a_0-a_4	a_0-a_2	0.70	2.13	0.33	1.73
5	(01,14,28)	a_0-a_2	a_0-a_2	1.48	4.48	0.27	1.34

Tab. 8.5: Clock prediction error at 100 min in nanoseconds with GIOVE clocks using different strategies

'dt' the fitting intervals, column 'Estimation' the parameters computed in the fitting and column 'BRD' the parameters to be broadcast to the user.

Several conclusions can be extracted from the results. The main improvement with respect to the basic strategy (#1) is obtained due to the reduction of the fitting interval (#2). Inclusion of the harmonic terms (#3) improves slightly the prediction at 100 minutes, the improvement being better at longer intervals. The harmonic term mainly helps to stabilize the fitting error as the provision of the additional coefficients to the user has no effect on the final error solutions, #3 and #4 being identical. Finally, the simple approach to use an integer multiple of the orbit period for the fitting interval (#5) provides the best accuracy for the PHM, while the error for the RAFS increases. This result is mainly due to the different σ_y at 6 and 14 hours for each clock. Obviously, the strategy which best suits one clock technology or unit may not be the best for another.

Stochastic model

The clock prediction objectives were twofold: first to reduce the prediction error in terms of standard deviation and maximum error, and second to assign a stochastic model to the clock prediction. An additional experiment is required for this second objective.

Equally important to have a good prediction is the possibility to associate a stochastic model to this prediction which can be used to provide a variance when computing the least squares adjustment. The stochastic model has been quantified following Equation 8.3:

$$\sigma_{x_p}(t) = \sqrt{\sigma_x^2 + \sigma_{a_0}^2 + (\sigma_{a_1}t)^2 + (\sigma_{a_2}t^2)^2 + (\sigma_{yWF}(t)t)^2 + (\tfrac{\sigma_{yFF}(t)t^2}{\ln 2})^2} \qquad [8.3]$$

where :

- $\sigma_{x_p}(t)$, is the expected clock error

- σ_x is the clock phase estimation error computed from the 1-sigma distribution of the different estimation arcs. GIOVE estimation processing runs every hour estimating clocks and orbits with the last 48 hours of data. As a consequence, 48 different clock samples are available for the same instant. The average value obtained is 0.3 ns (1σ).

- $\sigma_{a_0}, \sigma_{a_1}, \sigma_{a_2}$ are the a posteriori sigma of the least squares adjustment. The theoretical model for the clock prediction error is the one described in Equation 8.2. Such a formula is correct under the hypothesis of independent estimates of fit coefficients. If the coefficient estimates are not independent some correlation terms appear and have to be taken into account in the uncertainty estimation. In order to eliminate such terms or to have at least negative correlations (which would not be a problem in the worst case analysis) baricentric coordinates have to be used in the polynomial fit estimate.

- $\sigma_y^2(t)$ is the Allan variance of the clock evaluated at the time of prediction t. The stochastic contribution on the uncertainty on clock prediction has been evaluated considering two types of noise: white noise and flicker noise. For the PHM on GIOVE-B the values of such noises have been taken from the specifications previously covered in Section 4.2 (1E-12 for WFN and 1E-14 for FFN). For GIOVE-A the scecified value of flicker noise (3E-14) has been taken while for the white noise an experimental value of 6E-12 has been considered, which is bigger than the value reported in the specs (5E-12).

The stochastic model defined in Equation 8.3 has been applied and the expected error named as e. Additionally, as the fitting a posteriori sigma ($\sigma_{a_0}, \sigma_{a_1}, \sigma_{a_2}$) could not be representative of the adjustment an alternative approach has also been tested. The clock parameters a_0, a_1, a_2 estimated over the moving window $t_0 - 24h$ are stored at each $t0$ and used to compute alternative sigma values ($\sigma'_{a_0}, \sigma'_{a_1}, \sigma'_{a_2}$) by computing their standard deviation over the fitting interval. Obviously, a converge period of 1 day is required to obtain the first full set of a_{0-2} values over the moving window. These alternative sigmas ($\sigma'_{a_0}, \sigma'_{a_1}, \sigma'_{a_2}$) are used in Equation 8.3 instead of ($\sigma_{a_0}, \sigma_{a_1}, \sigma_{a_2}$) and the expected error named as e'.

The grey lines in Figure 8.12 represent the instantaneous error for every single prediction. The root mean square (rms) and the standard deviation (std) for the prediction error are computed at each prediction time. Both values present a good overlap indicating a zero mean unbiased distribution. The modified model e' follows with a better agreement the standard deviation (std) of the prediction for both strategies #2 and #5. On the contrary, the theoretical stochastic model e seems to underestimate the real error diverging for prediction times over 6 hours for strategy #2. This is most likely due to optimistic sigma values for the clock parameters (a_0, a_1, a_2) without taking into account the orbit period. It has to be also remarked how strategy #5 considerably reduces the error at 1 day.

In order to translate the clock predictions for all GNSS satellites, a strategy has to be selected. The inclusion of the harmonic term could make the fitting unstable in the case that no harmonic exists. Strategy 2 is applied to the complete constellation of GNSS satellites for the three years (selected operation period from 2008 till 2011). A prediction at 100 minutes is selected to

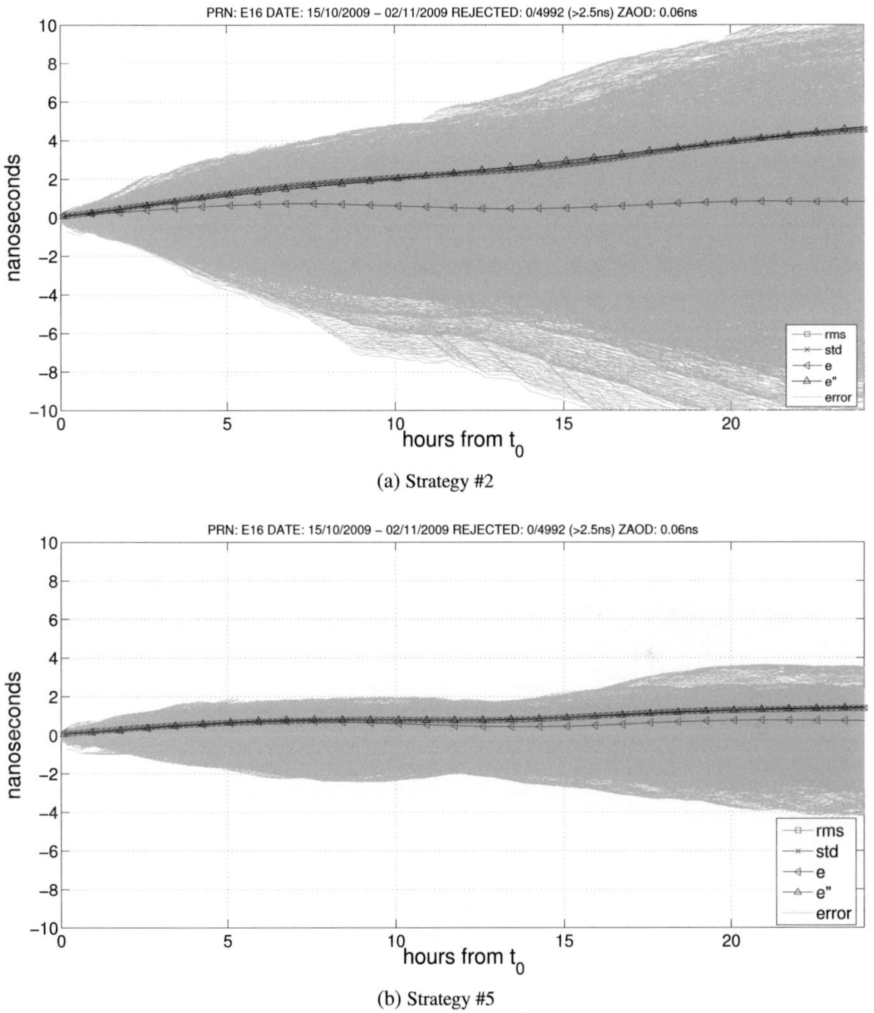

(a) Strategy #2

(b) Strategy #5

Fig. 8.12: Clock prediction for GIOVE-B in PHM mode over 24 hours

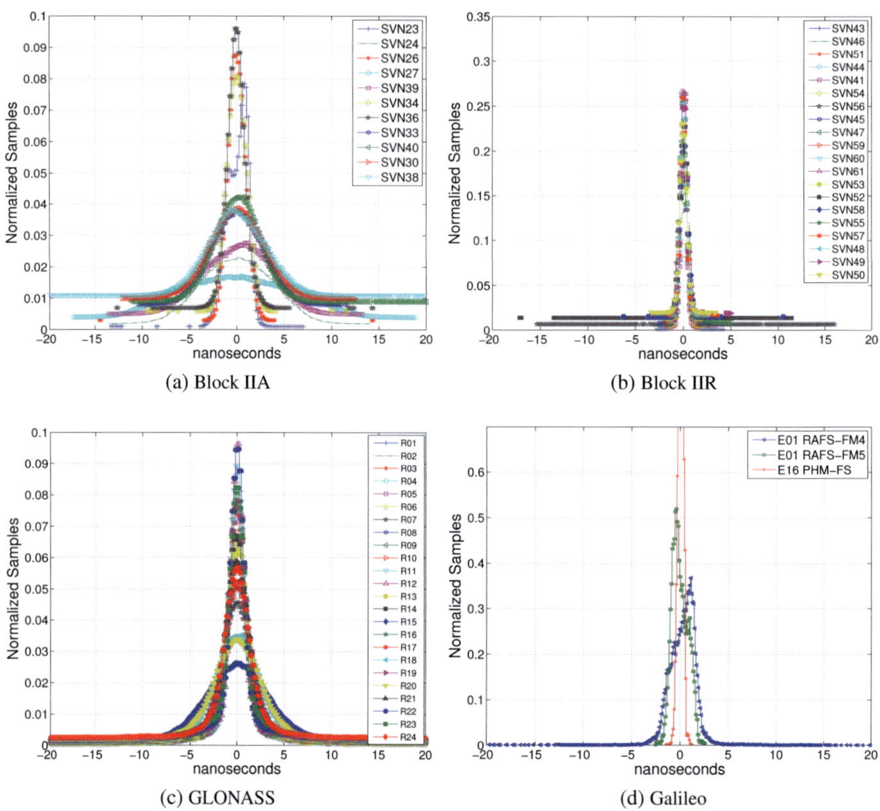

Fig. 8.13: Probability Density Function of GNSS clock prediction at 100 min

compare the clocks. IGS final clock estimations are used for GPS satellites. IGS-ESA analysis center estimations are used for GLONASS.

Figure 8.13 presents the probability density function (PDF) for each satellite according to the constellation type. The subfigures are normalized. The Y-axis scale must be carefully observed to get the indication of the distribution. Block IIA represents a mixed constellation of cesium and rubidium AFS with two types of distributions. Block-IIR shows a homogeneous constellation. All satellites have centred and symmetric distributions with the exception of SVN23. Galileo RAFS present asymmetric distributions most likely due to the non-monotonic frequency drift, while the PHM provides the best performance.

The single PDFs are slightly vertically displaced for each satellite to get a clear view of the tails. One of the more interesting features in Figure 8.13 is the longer tails observed for some satellites. The computation of the one sigma value is not affected by the tails due to the lower probability. The computed 1-sigma value cannot be associated to a normal distribution used in the stochastic models for navigation and safety of life applications requiring a high level of integrity, since the a-priori sigma will not cover these longer tails. A different methodology is required to provide a stochastic model for integrity users.

8.5 Integrity

The requirements for integrity in GNSS and SBAS originate from the International Civil Aviation Organization (ICAO) which defines the requirements for signal integrity, reliability, availability, and accuracy for the GNSS radio navigation aids used in civil aviation [72]. Three types of major error sources can be defined for the GNSS user linked to :

1. signal and navigation message generation (system errors).

2. signal propagation from transmitting to receiving antenna (environmental errors such as ionosphere, troposphere, multipath, interference)

3. signal processing by the user receiver (receiver errors).

Galileo tries to protect users by providing additional information about the system contribution (type-1), while SBAS includes additional ionosphere protection (type-2) for single frequency users. The remaining error source contributions are left to the user. The actual use of GNSS for positioning frequently lacks a rigorous stochastic model linked to the deterministic model.The variance associated with the observations and corrections is not provided nor are empirical fixed values used. In order to provide an integrity service to the user, a stochastic model needs to be defined that is linked to the orbit and clock predictions. The stochastic model can be simplified if Gaussian zero mean distributions are assigned to the observations and variance propagation laws are applied to the deterministic model. In this sense, in the ICAO standards, the PDF of the error shall be bound by a Gaussian PDF with a higher sigma. Once the variance linked to the pseudorange is provided, users may compute integrity figures according to their needs, as integrity risk or protection levels.

In practice, the bounding is practically performed based on the Cumulative Density Function (CDF) following for example [37]. The CDF is defined as the integral of the PDF:

$$CDF(X) = \int_{-\infty}^{X} \mathrm{pdf}(x)\mathrm{d}x = P(x \leq X) \qquad [8.4]$$

Once the CDF of an experimental random variable A is computed, it is considered overbound by a normal distribution B, if its distribution is smaller than the normal distribution B for any error interval.

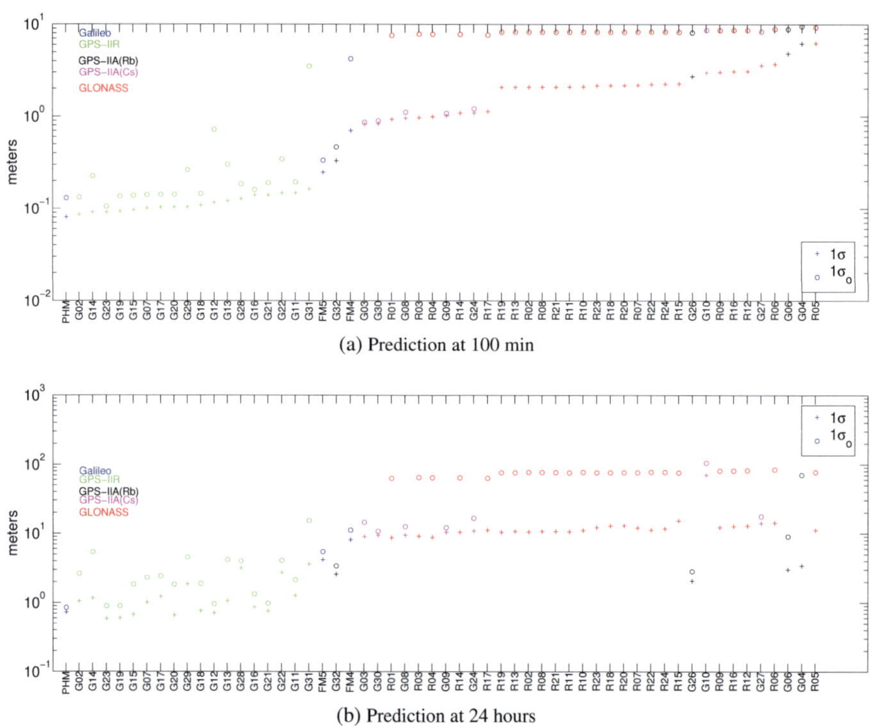

(a) Prediction at 100 min

(b) Prediction at 24 hours

Fig. 8.14: Clock prediction with overbounding (in metrical units)

$$CDF_A(X) \leq CDF_B(X) \quad \forall X \qquad [8.5]$$

In reality, the overbounding B distribution is computed by selecting the first normal distribution which is above the CDF function for A for any interval. The CDF over-bounding, defined in [37] cannot work for both tails separately at the same time when the experimental distribution has a bias. The Galileo integrity concept explained in [121] introduces a slight modification by combining both tails to overbound the absolute CDF.

$$CDF_A(|X|) \leq CDF_B(|X|) \quad \forall X \geq 0 \qquad [8.6]$$

The principal difficulties arise in overbounding the real distribution with a zero mean Gaussian distribution, since biases, asymmetries and large tails can be associated to the distribution. Galileo and WAAS solution is to overbound the core distribution using a mathematical method assigning probabilities to the tails of the distribution through analysis of possible error sources (e.g.probability of a clock failure). This approach limits the worst case behaviour of the clock errors outside the distribution, and allocates a certain probability to any event. Assigning probabilities for a complete constellation with different clock technologies and even single clock behaviour is not an easy task. Using a conservative approach may invalidate the complete integrity concept.

The application of the overbounding concept is shown in Figure 8.14 where the 1σ distribution obtained in the previous section has been overbounded with the $1\sigma_o$ distribution. Satellites are ranked from the lowest to the highest 1σ for the 100 minutes prediction. This order is maintained for the prediction at 24 hours. The colour identifies the timing subsystem family. Some of the satellites present significant differences between the 1σ and the associated overbounded $1\sigma_o$ value. This difference is due to the overbounding required to include the tails within the new normalized distribution. A significant increase is required for several Block IIA and all GLONASS satellites. Conversely, the best agreement between both sigmas is observed for the PHM and Block-IIR families.

8.6 Conclusions

This chapter has started highlighting the importance of the AFS stability in GNSS satellites in providing an accurate clock prediction to the user. The stability of current GNSS AFS has been investigated with a focused attention on the Galileo clocks family. It has been concluded that the high frequency stability and low drift make the PHM the best operating GNSS clock in terms of performance, providing new possibilities for navigation.

In the following, it was realized how frequency steps and maintenance operations may have an impact on the clock time scale. The clock prediction strategy for real-time systems should be protected against these kinds of events during the fitting interval. Events before and after the end of the fitting interval will affect the tails of the distribution function. Suggestions on how to implement a robust clock prediction strategy, able to cope with these events, have been provided.

This section has highlighted how the harmonics in GNSS clocks hamper the prediction. To resolve this problem, it has been demonstrated how the inclusion of harmonic coefficients in the prediction increases the overall accuracy. Furthermore, the improvement in the fitting makes no longer necessary to transmit the additional harmonic coefficients to the user but just the first polynomial coefficients. This concept can be used by current GNSS systems with a legacy message. In PHM mode, the simple selection of a fitting interval proportional to the orbit period provides the best results. Additionally, an associated stochastic model to the prediction has been proposed with good agreement up to 1 day to the a-posteriori observed real predictioon error.

In the introduction it was identified how the predictions generally lack an associated stochastic model. In this chapter, a stochastic model to be linked to the prediction has been proposed and validated using a reference period. Additionally, some users require a robust prediction.

The integrity methodology used in civil aviation has been applied to all GNSS clocks. Their probability density function distributions may have associated long tails which require a large 1-sigma increase in order to prevent erroneous information getting to the user. The sigma is increased by selecting the standard deviation of the normal distribution whose cumulative distribution function is above the observed one. As a consequence these satellites will have lower weight in a PVT solution with respect to other more robust satellites. By technology the PHM provides the most robust signal followed by Block IIF satellites (rubidium+TKS) with only 2 satellites out of 20 with significant increases. The rest of the AFS families require significant inflation of the 1-sigma distribution.

9 Conclusion and outlook

The main subject of this thesis is the new 'clocks' and 'timing signals' available in the new navigation satellites part of Galileo, GPS Block-IIF, GLONASS-K and COMPASS.

On the background of the new complexity associated with the new frequencies, modulations and systems, one of the central questions is a clarification of the relationship between the different 'clock concepts' currently being used. The Atomic Frequency Standard (AFS) specifications are broadly used to derive the expected performance and prediction accuracy of the new satellite time scales (t^{sat}).This dissertation has demonstrated how this assumption can lead to erroneous conclusions.

The name 'clock' is usually applied to the atomic frequency standard (AFS) on board a satellite even if it does not directly provide time information. The AFS generates a reference frequency f_i converted to F_0 by a frequency control unit which further provides the signal to a navigation unit, where the 'Timing Signal' is physically created by the encoding of the navigation codes and time-tag information. This timing signal is filtered, amplified and broadcast to the user by other parts of the payload. This 'timing signal' is recovered by the receiver in terms of phase and code measurements. Traceability from ground to satellite time is performed using ionosphere-free combinations of two timing signals and grouping hardware delays into the estimated clocks . Based on this information, this work clarifies that a clear separation in three different concepts is more appropriated when referring to the on-board 'clock':

physical clock : the Atomic Frequency Standard generating the basic frequency F_0.

signal clock : the navigation signal at the output of the satellite antenna. It is independent for each specific signal and modulation, as each one includes different hardware delays.

ionosphere-free clock : or apparent clock observed on ground by Precise Orbit Determination techniques based on a pair of frequencies.

In order to provide the time signal offset to the user, Global Navigation Satellite Systems (GNSS) compute the position and traceability between the satellite time and the system time $t^{sat} - t_{sys}$ by geodetic time transfer methodology. This thesis reviewed the methodology and accuracy achieved by geodetic time transfer in order to identify the limits and possible improvements. The review of the methodology revealed that the satellite position estimation makes time estimations strongly correlated with the orbit. The expected theoretical one-way time transfer limit was here demonstrated to be 100 ps (1σ) accurate from code and 1 ps (1σ) precise from carrier phase measurements. The weighting scheme derived from this 1/100 factor provides the time transfer accuracy based on the code and the precision on the ambiguous carrier phase observations. In practice, the state of the art of geodetic time transfer achieves, for clock products, 0.07 ns (rms), 20 ps (1σ) and $1\text{E-}12\tau^{-1/2}$ frequency stability. Nevertheless, typical values of 1 ns (rms) accuracy were also observed, while comparison with respect to an independent time transfer technique (Two-Way Satellite Time and Frequency Transfer) was demonstrated to be consistent at 2 ns.

Before the second Galileo In Orbit Validation Element (GIOVE-B) launch in 2007, it was already clear that the performance of the new Passive Hydrogen Maser (PHM) frequency standard would be at the limit of the state of the art of geodetic time transfer capabilities and above the capabilities of the envisaged ground segment. Due to the limited number of stations the geodetic time transfer performance achieved by GIOVE mission is obtained here to be 0.5 ns (rms), 0.3 ns (1σ) and $2.2\text{E-}12\tau^{-1/2}$ frequency stability. This observed stability is twice as inaccurate as the initially anticipated $1\text{E-}12\tau^{-1/2}$ value for the PHM and at the level of the best performing Rubidium Atomic Frequency Standard (RAFS) in GIOVE satellites. The first objective of the GIOVE mission was the reservation of the frequencies allocated by the International Telecommunication Union. The second goal was the validation of the payload equipment to be flown in Galileo; due to the fact that no European atomic clock was previously launched into space, their validation was the subsequent main objective of the GIOVE mission. Another methodology was required to verify the in-orbit performance of GIOVE clocks.

A novel methodology was proposed within this thesis, which was described, implemented in a dedicated software and then validated with GPS satellites with an excellent agreement against International GNSS service (IGS) results. The short term behaviour below 300 seconds is not covered by IGS final products. The combination of this methodology and POD results additionally allowed the characterization of GNSS clocks from 1 second for the first time. Once confirmed its suitability to characterize GNSS clocks, it was applied to GIOVE clocks. It has been proven how the short term stability of RAFS and PHM are in line with the ground measurements; it is even possible to identify the activated RAFS unit from the agreement. This agreement has validated this novel methodology, which has allowed the first full characterization of GNSS clocks and the successful achievement of the second objective of the GIOVE mission by validating the PHM and RAFS clock performance. This new methodology has been further adopted and customized by other groups such as the French or German Aerospace Centres [39, 112, 69] to achieve similar results.

The only unknown effect observed in the PHM is a 0.5 ns harmonic in the estimated ionosphere-free clock. While harmonics in GPS satellites have been a well-known feature since their early identification [158], only one recent publication [150] has briefly mentioned the temperature as the origin of this effect but it still lacks any dedicated analysis. The origin of the harmonic in the apparent clock of GNSS satellites is reviewed and clarified in this dissertation. In this dissertation, it has been demonstrated how the amplitude correlates with the sun-beta angle for most of the satellites. This correlation indicates a possible dependency on temperature. A simple methodology has been proposed to derive the expected harmonic from the sensitivity of the physical clocks with respect to temperature. The agreement observed between expected and measured values indicates that harmonics in the apparent clock of GNSS satellites are mainly due to the thermal sensitivity of the AFS. The only disagreement is observed for the PHM on-board GIOVE-B, where temperature-induced variations in the AFS seem unlikely in view of several indications. The harmonic was already predicted before the satellite launch as an artificial effect due to the orbit accuracy possible with the envisaged 13 stations. This hypothesis has been demonstrated by the reduced amplitude when the number of measurements is increased by adding stations, satellite laser ranging measurements or by extending the arc length.

The orbit period component indicates the Solar Radiation Pressure (SRP) model as a probable cause of the harmonic in the PHM estimation. The empirical model used for SRP estimation may be inaccurate or affected by the degraded geometry due to the low number of stations. Once the Galileo constellation is deployed, the accuracy of the SRP for Galileo satellites in PHM mode should be reviewed at the time when a higher number of sensor stations becomes available.

The special attention given here to the harmonic is not unimportant; this effect impacts the clock prediction which then impacts the user. Clock offset $dt^s(t)$ prediction still represents one of the major error contributors for real time navigation and the main limitation for extended ephemeris use. Clock corrections are also the main added value of real time double frequency based services. The harmonic origin should be understood towards the implementation of possible mitigation strategies at system or user level.

Independently from the origin of the harmonics, the inclusion of harmonic coefficients in the ionosphere-free clock prediction has been demonstrated here to increase the accuracy. The gain is mainly observed in the polynomial terms making it not necessary to transmit the harmonic coefficients to the user. In PHM mode the prediction error is at the same level as the estimated noise (0.3 ns,1σ) at 100 min and at the level of the harmonics (0.5 ns) attributed to the orbit at 1 day. Independent of the prediction strategy, a stochastic model has been proposed for the prediction, with an excellent agreement with the real error.

In terms of scientific advance, the superior frequency repeatability of the new clock technology provided by the PHM has allowed us to measure the expected 4.718E-10 relativistic frequency shift to within an error of 5.58E-12, corresponding to 1.2% of the measured value. Additionally, the currently applied periodic relativistic correction has a periodic error of 0.1 ns, as announced by Kouba [83] - whilst this effect is hidden in other GNSS clocks, it is clearly visible with the PHM.

All these facts demonstrate how the new PHM has brought the physical clock error contribution below the noise floor of geodetic time transfer capabilities. The physical clock characteristics are not observed in the ionosphere-free clock, while in the case of other technologies (RAFS and cesium) their contribution is dominant. In terms of prediction, the PHM frequency stability (falling up to 2.5E-15) makes it possible to completely cancel the clock error contribution in the navigation solution and render any real time overlay service redundant. In timekeeping, the low frequency drift (7E-16 $df/$day) makes any constraints concerning the steering of satellite clock time practically dissapear. While GNSS constellations are slowly evolving, the new generation of optical clocks is being developed on the ground, promising a better level of performance (down to the 1E-18 level). It seems that the limitations in geodetic time transfer identified in this dissertation with the PHM will have to be addressed before using the full potential of these enhanced upcoming clocks. Although optical clocks are not expected to be available in the medium term for GNSS systems, they will generate a major interest in the research field. It is strongly advised to introduce this technology in a fundamental physics space mission [49], opening the way for their future use in GNSS payloads.

Future perspectives

One of the central questions of this thesis was to assess the new opportunities brought by the new GNSS clocks. This dissertation has proposed new methodologies for satellite clock characterization, accurately measured the relativistic net frequency shift, demonstrated the presence of second order periodic relativistic contributions and proposed a novel methodology to quantify the harmonic contributions. These results represent a major step forward giving some directions for future systems to robustly predict or use clock prediction depending on their needs.

However, there are still some aspects related to the new AFS possibilities that have been identified but not addressed in this thesis.

1. Scientific aspects:

    a) It will be possible to refine the measurement of the net relativistic frequency shift once the semi-major axis of the Galileo satellite orbit is increased to the graveyard orbit, or by traceability of the H-maser used in ground tests to a reference time scale.

b) Based on more accurate satellite and receiver frequency sources, the possibility of using a dynamic model for the clock (based on physical parameters) would considerably reduce the amount of unknowns/variables and the correlation with other estimated parameters. PHM on-board Galileo satellites and H-masers at ground stations can already provide a validation of this concept. Nonetheless, the group delays currently included in the 'iono-free clock' have been identified as a potential disturbing component and must be carefully taken into account.

c) The observed harmonic period is not exactly the value of the orbit period. This dependency should be analysed and the difference in the harmonic period further investigated with 30 seconds products and more Galileo PHMs.

d) The SRP coefficients are dependent on the argument of latitude and as a consequence of the orbit period. The empirical model used for SRP estimation may be inaccurate. The low noise of PHM clocks would allow the review of its accuracy once a higher number of sensor stations are available to track these satellites.

e) The low clock noise provided by the PHM allows clear observation of the attitude mis-modelling during the midnoon turns. This effect may be used to search for the optimum attitude model.

f) The evolution of the current double-frequency approach used in Precise Orbit Determination to multi-frequency, by estimation of the ionosphere contribution based on the numerous frequencies and systems available, could open new possibilities for ionosphere, orbit, clock and differential code bias products.

2. System management:

a) Indications have been provided for a robust clock prediction strategy. The use of flexible intervals depending on the clock model and unit behaviour seems to be the best option. This hypothesis could be studied in more depth to define a better strategy than a generic one for all units.

b) The Frequency Distribution Unit in the satellite allows the implementation of capabilities for autonomous detection of clock anomalies, in order to increase the reliability and integrity of the timing subsystem. This approach is being reviewed for future Galileo satellites and has already been implemented by the Time Keeping System (TKS) in GPS Block-IIR. In the near future, a single robust AFS seems the most convenient and simplest strategy for navigation satellites as demonstrated by the re-introduction of a similar FDU design from GPS Block IIA in Block IIF. Nevertheless, the clock data collected from GIOVE satellites could be used to define and test a prototype algorithm.

c) The newly developed methodology to extract the short term noise could be used in combination with normal network estimations to derive the full clock performance for each clock family, and estimate the optimum interpolation, prediction and sampling time for each clock family.

d) The current tendency for high precision applications is to develop real time services to provide the clock model at regular intervals with a high repetition rate at regular intervals. This approach requires a large bandwidth. Based on the stability of GNSS orbits and recently available clocks (such as PHM) a lower latency and flexible update rate depending on the clock performance may allow the provision of the same performance but with a much lower bandwidth.

3. Mass-market users: Ephemerides extension techniques used for decreasing the time to first fix by commercial receivers are limited to a few days due to the accuracy of existing clocks. The PHM has been demonstrated to remain within a ± 20 ns range when 6 months of data are detrended. It should be possible to review current methodologies to significantly increase the extension period.

10 Bibliography

[1] J. Achkar, P. Tuckey, P. Ullrich, D. Valat, A. Batchelor, G. Burden, A. Bauch, D. Piester, F. Cordara, P. Tavella, et al. Fidelity-progress report on delivering the prototype galileo time service provider. In *Frequency Control Symposium, 2007 Joint with the 21st European Frequency and Time Forum (EFTF). IEEE International*, pages 446–451. IEEE, 2007. 18

[2] C. Affolderbach, R. Matthey, F. Gruet, T. Bandi, and G. Mileti. Realisation of a compact laser-pumped rubidium frequency standard with < 1x10-12 stability at 1 second. In *24th European Frequency and Time Forum (EFTF)*, (ESA/ESTEC) Noordwijk, The Netherlands, April 2010. 59

[3] D. Allan. Time and frequency (time-domain) characterization, estimation, and prediction of precision clocks and oscillators. *Ultrasonics, Ferroelectrics and Frequency Control, IEEE Transactions on*, 34(6):647–654, 1987. 119

[4] D. Allan and M. Weiss. Accurate time and frequency transfer during common-view of a gps satellite. In *Proc. 34th Ann. Freq. Control Symposium*, USAERADCOM, Ft. Monmouth, WJ 07703, USA, May 1980. 74

[5] M. Anghileri and M. Paonni. Ready to navigate! a methodology for the estimation of the time-to-first-fix. *Inside GNSS*, pages 47–56, March 2010. 22

[6] M. Anghileri, M. Paonni, S. Wallner, J.-A. Avila-Rodríguez, and B. Eissfeller. A methodology for the estimation of the time-to-first-fix. *Inside GNSS*, April 2010. 20, 22

[7] N. Ashby. Practical implications of relativity for a global coordinate time scale. *Radio Science*, 14(4):649–669, July-August 1979. 36

[8] N. Ashby. Relativity in the future of engineering. *IEEE Transactions on instrumentation and measurement*, 43(4):505–514, August 1994. 36

[9] N. Ashby. Relativity in the global positioning system. *Living Reviews in Relativity*, 6(1), 2003. [Online Article]: cited in June-2012. 36, 41

[10] C. Audoin and B. Guinot. *The Measurement of Time*. Cambridge University press, 1 edition, 1998. 50

[11] A. Baker. GPS Block IIR Time Standard Assembly (TSA) Architecture. In *22nd Precise Time and Time Interval (PTTI) Meeting*, pages 317–324, Vienna, Virginia,USA, 1990. 63, 142

[12] Y. Bar-Sever and D. Kuang. New empirically derived solar radiation pressure model for global positioning system satellites. *IPN Progress Report*, 42:159, 2004. 152

[13] B. Bartenev, V. Kosenko, and V. Cheboraterev. Russian GLONASS at the stage of active implementation. *Inside GNSS*, April 2006. 144

[14] V. Bartenev, V. Kosenko, and V. Chebotarev. Builders notes: Russian GLONASS at the stage of active implementation. *Inside GNSS*, April 2006. 141

[15] A. Bassevich and P. P. Bogdanov. GLONASS onboard time/frequency standards: ten years of operation. In *28th Annual Precise Time and Time Interval (PTTI) Meeting*, pages 455–462, Reston,Virginia,USA, December 1996. 54

[16] A. Bassevich, B. Shebshaevich, A. Tyulyakov, and V. Zholnerov. Onboard atomic clocks GLONASS: current status and future plans (GLONASS Modernization, QZSS and Other GNSS). In *International Technical Meeting of the Satellite Division of the Institute of Navigation (ION GNSS)*, Fort Worth, Texas,USA, September 2007. Power point presentation. 54, 59

[17] S. Bassiri and G. Hajj. Higher-order ionospheric effects on the gps observables and means of modeling them. *Manuscripta Geodaetica*, 18:280–289, 1993. 86

[18] A. Bauch. Caesium atomic clocks: function, performance and applications. *Measurement Science and Technology*, 14(8):1159–1173, 2003. 52, 53, 142

[19] M. Belloni, M. Gioia, and S. Beretta. Space passive hydrogen maser-performances, lifetime data and GIOVE-B related telemetries. In *24th European Frequency and Time Forum (EFTF)*, pages 393–398, ESA-ESTEC,Noordwijk,NL, April 2010. 59, 142

[20] M. Belloni, M. Gioia, S. Beretta, F. Droz, P. Mosset, and P. Waller. Space mini passive hydrogen maser-a compact passive hydrogen maser for space applications. In *Frequency Control and the European Frequency and Time Forum (EFTF), 2011 Joint Conference of the IEEE International*, pages 1–5. IEEE, 2011. 61

[21] J. Benedicto, S. E. Dinwiddy, G. Gatti, R. Lucas, and M. Lugert. Galileo: Satellite system design and technology developments. Technical report, ESA, Nov 2000. 65

[22] J. W. Betz and K. R. Kolodziejski. Generalized theory of code tracking with an early-late discriminator part II - noncoherent processing and numerical results. *Aerospace and Electronic Systems, IEEE Transactions on*, 45(4):1557–1564, October 2009. 95

[23] G. Beutler, E. Brockmann, W. Gurtner, U. Hugentobler, L. Mervart, M. Rothacher, and A. Verdun. Extended orbit modeling techniques at the code processing center of the international gps service for geodynamics (igs): theory and initial results. *Manuscr. Geod., Vol. 19, No. 6, p. 367-386*, 19:367–386, 1994. 150

[24] BIPM. *The International System of Units (SI)*. STEDI MEDIA, 8 edition, 2006. 14, 35, 49, 50

[25] K. Borre. *Plane networks and their applications*. Springer, 2001. 79

[26] C. Bourga, B. Lobert, and M. Brunet. Characterisation of Galileo Clocks Onboard an Experimental Satellite. In *17th European Frequency and Time Forum (EFTF)*, St-Petersburg, Russia, 2002. 147

[27] R. Braun. GIOVE experimentation results for Galileo system performance simulations. In *International Technical Meeting of the Satellite Division of the Institute of Navigation (ION GNSS)*, Portland, Oregon,USA, September 2010. 113

[28] R. V. Bree, C. Tiberius, and A. Hauschild. Real time satellite clock corrections in precise point positioning. In *International Technical Meeting of the Satellite Division of the Institute of Navigation (ION GNSS)*, pages 22–24, Georgia, USA, May 2008. 173

[29] G. Burbidge. Development of the navigation payload for the Galileo in-orbit validation (IOV) phase. In *International Global Navigation Satellite Systems Society IGNSS Symposium 2007*, The University of New South Wales, Sydney, Australia, December 2007. 20

[30] L. Cacciapuoti, N. Dimarcq, G. Santarelli, P. Laurent, P. Lemonde, A. Clairon, P. Berthoud, A. Jornod, F. Reina, S. Feltham, et al. Atomic clock ensemble in space: scientific objectives and mission status. *Nuclear Physics B-Proceedings Supplements*, 166:303–306, 2007. 76

[31] A. Cartmell. Considerations for calibration of frequency dependent delays. In *International Technical Meeting of the Satellite Division of the Institute of Navigation (ION GNSS)*, pages 799–809, Salt Lake City, USA, September 2000. CAA Institute of Satellite Navigation, University of Leeds, UK. 67

[32] G. Cerretto, N. Guyennon, I. Sesia, P. Tavella, F. Gonzales, J. Hahn, V. Fernandez, and A. Mozo. Evaluation of the odts system noise in the galileo giove mission. In *22nd European Frequency and Time Forum (EFTF)*, Toulouse,France, April 2008. 106

[33] G. Cerretto, A. Perucca, P. Tavella, A. Mozo, R. Piriz, and M. Romay. Network time and frequency transfer with gnss receivers located in time laboratories. *Ultrasonics, Ferroelectrics and Frequency Control, IEEE Transactions on*, 57(6):1276–1284, 2010. 75

[34] Y. C. Chan, J. C. Camparo, and R. P. Frueholz. The autonomous detection of clock problems in satellite timekeeping systems. In *31st Precise Time and Time Interval (PTTI) Meeting*, pages 111–120, Dana Point, California,USA, 1999. 65

[35] R. Dach, U. Hugentobler, P. Fridez, and M. Meindl. Bernese gps software version 5.0 (user manual of the bernese gps software version 5.0). Technical report, University of Bern, 2007. 78

[36] T. Dass, J. Petzinger, J. Rajan, and H. Rawicz. Analysis of on-orbit behavior of GPS BLOCK II-R TKS. In *31st Precise Time and Time Interval (PTTI) Meeting*, pages 1973–1986, Dana Point, CA,USA, 1999. 64

[37] B. DeCleene. Defining pseudorange integrity overbounding. In *International Technical Meeting of the Satellite Division of the Institute of Navigation (ION GPS)*, pages 1916–1924, Salt Lake City, Utah,USA, 2000. 179, 180

[38] P. Defraigne and C. Bruyninx. On the link between gps pseudorange noise and day-boundary discontinuities in geodetic time transfer solutions. *GPS Solutions*, 11:239–249, 2007. 10.1007/s10291-007-0054-z. 103, 106

[39] J. Delporte, C. Boulanger, and F. Mercier. Straightforward estimations of GNSS on-board clocks. In *Frequency Control and the European Frequency and Time Forum (EFTF), 2011 Joint Conference of the IEEE International*, pages 1–4. IEEE, 2011. iii, 184

[40] G. L. Dieter. Observations on the reliability of Rb FS on board block IIA satellites. In *27th Annual Precise Time and Time Interval (PTTI) Meeting*, pages 125–134, Falcon Air Force Base Colorado Springs, CO, USA, November 1995. 56, 57

[41] F. V. Diggelen. *A-GPS: assisted GPS, GNSS, and SBAS*. Artech House Publishers, 2009. 20

[42] DoD. GPS standard positioning service performance standard. Technical report, USA Department of Defense, September 2008. 169, 172, 173

[43] F. Droz, P. Rochat, and Q. Wang. Performance overview of space rubidium standards. In *24th European Frequency and Time Forum (EFTF)*, (ESA/ESTEC) Noordwijk, The Netherlands, April 2010. 58

[44] R. T. Dupuis, T. J. Lynch, and J. R. Vaccaro. Rubidium frequency standard for the GPS IIF program and modifications for the RAFSMOD Program. In *Frequency Control Symposium, 2008 IEEE International*, pages 655–660, Salem, MA 01970 USA, November 2008. IEEE. 54, 57, 144

[45] V. Engelsberg, V. Babakov, and I. G. Petrovski. Expert advice - GLONASS business prospects. *GPS World*, March 2008. http://www.gpsworld.com/gnss-system/glonass/expert-advice-glonass-business-prospects-4215. 141

[46] M. Epstein and T. Dass. Management of phase and frequency for GPS IIR satellites. In *33rd Annual Precise Time and Time Interval (PTTI) Meeting*, pages 481–492, The University of New South Wales, Sydney, Australia, November 2001. 25

[47] M. Epstein, T. Dassa, J. Rajan, and P. Gilmour. Long term clock behaviour of GPS-IIR satellites. In *39th Ann. Precise Time and Time Interval (PTTI) Meeting*, Long Beach, California,USA, November 2007. 57, 162

[48] ESA. *GIOVE-A+B navigation signal in space interface control document*, 1.1 edition, Augustus 2008. 16

[49] ESA. A roadmap for fundamental physics in space. Technical report, ESA Fundamental Physics Roadmap Advisory Team, 2010. v, 62, 186

[50] ESA. *GIOVE Experimentation Results - A Success Story*. ESA Communications, November 2011. 5, 142, 150

[51] L. Essen. Time scales. *Metrologia*, 4(4):161–165, 1968. 14, 49

[52] European Union. *European GNSS (Galileo) open service signal in space interface control document*, 1.0 edition, February 2010. 16

[53] D. Felbach, F. Soualle, L. Stopfkuchen, and A. Zenzinger. Clock monitoring and control units for navigation satellites. In *Frequency Control Symposium (FCS), 2010 IEEE International*, pages 474–479. Astrium GmbH, 2010. 64

[54] D. Felbach, F. Soualle, L. Stopfkuchen, and A. Zenzinger. Future concepts for on-board timing subsystems for navigation satellites. In *24th European Frequency and Time Forum (EFTF)*, (ESA/ESTEC) Noordwijk, The Netherlands., 2010. Astrium GmbH. 65

[55] S. Fisher and K. Ghassemi. GPS IIF-the next generation. *Proceedings of the IEEE*, 87(1):24–47, 1999. 62, 63

[56] G. Galluzzo, M. Sánchez-Gestido, F. Gonzalez, S. Binda, G. Radice, A. Hedqvist, and R. Swinden. GIOVE-B navigation message performance analysis and signal in space user ranging error (SISRE) characterization. In *International Technical Meeting of the Satellite Division of the Institute of Navigation (ION GNSS)*, Savannah, Georgia, USA, September 2009. 173

[57] G.Gatti, M.Falcone, V.Alpe, T.Burger, M.Rapisarda, and E.Rooney. GIOVE-B Chilbolton in orbit test. *Inside GNSS*, pages 30–35, September 2008. 113

[58] A. Gifford, S. Pace, and J. McNeff. One-way gps time transfer 2000. In *32nd Annual Time and Time Interval (PTTI) Systems and Applications Meeting*, 2000. 74

[59] Global Positioning System Wing. *Navstar GPS Space Segment. User Segment L5 Interface, IS-GPS-705, Revision A*, June 2010. 16, 21

[60] Y. Gouzhva, A. Gevorkyan, A. Bassevich, P. Bogdanov, and A. Tyulyakov. Comparative analysis of parameters of GLONASS spaceborne frequency standards when used onboard and on service life tests. In *Frequency Control Symposium, 1993. 47th., Proceedings of the 1993 IEEE International*, pages 65–70. IEEE, 1993. 54

[61] Y. Gouzhva, A. Gevorkyan, A. Myasnikov, G. Kryukov, V. Gankin, O. Kornishov, and S. Teplova. Activities of the Russian institute of radionavigation and time in development of GLONASS on-board Q-enhanced oscillating compact H-maser. In *49th. Frequency Control Symposium*, pages 140–148. IEEE, 1995. 60

[62] GPS-World. PerkinElmer unit awarded $15 million GPS III clock contract. *GPS world*, September 2009. 57

[63] J. Griffiths and J. Ray. On the precision and accuracy of IGS orbits. *Journal of Geodesy*, 83(3):277–287, 2009. 95, 102

[64] W. Gurtner and L. Estey. *RINEX, The Receiver Independent Exchange Format*, November 2007. 31

[65] J. Hahn, F. Gonzalez, P. Waller, D. Navarro-Reyes, R. Piriz, V. Fernandez, P. Tavella, and I. Sesia. GIOVE-A apparent clock assessment and results. In *39th Annual Precise Time and Time Interval (PTTI) Meeting*, Long Beach, CA,USA, November 2007. 143

[66] D. Hanson. Fundamentals of two-way time transfers by satellite. In *Frequency Control, 1989., Proceedings of the 43rd Annual Symposium on*, pages 174–178. IEEE, 1989. 76

[67] A. Hauschild and O. Montenbruck. Kalman-filter-based gps clock estimation for near real-time positioning. *GPS Solutions*, 13:173–182, 2009. 10.1007/s10291-008-0110-3. 173

[68] A. Hauschild, O. Montenbruck, J.-M. Sleewaegen, L. Huisman, and P. Teunissen. Characterization of Compass M-1 signals. *GPS Solutions*, pages 1–10, 2011. 10.1007/s10291-011-0210-3. 58, 98

[69] A. Hauschild, O. Montenbruck, and P. Steigenberger. Short-term analysis of gnss clocks. *GPS Solutions*, 17(3):295–307, 2013. 184

[70] C. Hegarty, E. Powers, and B. Fonville. GPS + modernized GPS + Galileo signal timing biases. *GPS world*, March 2006. 65, 68

[71] V. Hermann, M. Gazard, P. Berthoud, S. Lecomte, R. Ruffieux, C. Audoin, R. Barillet, E. de Clercq, and S. Guerandel. Oscc project : A space cs beam optically pumped atomic clock for galileo. In *Frequency Control Symposium, 2007 Joint with the 21st European Frequency and Time Forum (EFTF). IEEE International*, pages 77 – 80. IEEE, 2007. 55

[72] ICAO. Standard and recommended procedures (SARPS),Annex 10. Volume I - Radio navigation aids. Technical report, ICAO, September 2008. 169, 179

[73] IERS. Chapter 10:general relativistic models for space-time coordinates and equations of motion. Working Version, unpublished, May 2009. 15, 16, 36, 40

[74] Indra. *Galileo Experimental Seensor Station specifications*, 2008. 8

[75] ISO. International vocabulary of basic and general terms in metrology. Technical Report 3.0, International Organization for Standarization, Geneva, 1993. 50

[76] H. Ito, T. Morikawa, S. Hama, M. Takizawa, Y. Numata, and S. Mattori. Development and performance evaluation of spaceborne hydrogen maser atomic clock in NICT. In *Institute of Navigation National Technical Meeting (ION NTM)*, pages 1–3, 2007. 60

[77] H. Ito, T. Morikawa, H. Ishida, S. Hama, K. Kimura, and S. Yokota. Development of a spaceborne hydrogen maser atomic clock for quasi-zenith satellites. In *36th Annual Precise Time and Time Interval (PTTI) Meeting*, pages 423–430, Washington. DC, USA, 2005. 60

[78] ITT. Evolution of the GPS navigation payload a historical journey. Technical report, Stanford Center for Position,Navigation and Time, June 2009. http://scpnt.stanford.edu/pnt/PNT09/presentation_slides/5_Brodie_History_GPS_Payload.pdf. 62

[79] ITU. A coordination methodology for RNSS inter-system interference estimation. Technical report, ITU, 2007. 98

[80] P. S. Jorgensen. Obtaining the short term stability of the GPS satellite clocks from tracking data. In *Institute of Navigation, National Technical Meeting*, pages 235–239, Long Beach, California,USA, January 1984. 119

[81] J. Kitching, S. Knappe, L. Liew, J. Moreland, H. Robinson, P. Schwindt, V. Shah, V. Gerginov, and L. Hollberg. Chip-scale atomic clocks at NIST. In *National Conference of Standards Laboratories (NCSL) International Workshop and Symposium*, Washington, DC,USA, 2005. 33

[82] F. Koide, D. Dederich, and L. Baker. Rubidium frequency standard test program for NAVSTAR GPS. In *10th Precise Time and Time Interval(PTTI) Meeting*, pages 379–392, Greenbelt, Maryland, USA, November 1978. 56

[83] J. Kouba. Improved relativistic transformations in GPS. *GPS Solutions*, 8:170–180, 2004. iv, 41, 186

[84] J. Kouba. A simplified yaw-attitude model for eclipsing gps satellites. *GPS Solutions*, 13(1):1–12, February 2009. 102

[85] J. Kouba and P. Héroux. Precise Point Positioning using IGS orbit and clock products. *GPS Solutions*, 9(2):12–28, July 2001. 105, 122

[86] J. Kouba and T. Springer. New IGS station and satellite clock combination. *GPS Solutions*, 4(4):31–36, 2001. 78, 85

[87] H. J. Kramer. GIOVE-B (Galileo In-Orbit Validation Element-B). Technical report, eoPortal, Jan 2008. http://directory.eoportal.org/presentations/182/10000377.html. 66

[88] R. B. Langley. Rubidium clock on SVN62: a bit disappointing. *GPS World*, August 2010. 145

[89] K. Larson and J. Levine. Time Transfer using the Phase of the GPS Carrier. In *Frequency Control Symposium, 1998. Proceedings of the 1998 IEEE International*, pages 292–297. IEEE, 1998. 119

[90] K. Larson and J. Levine. Carrier-phase time transfer. *Ultrasonics, Ferroelectrics and Frequency Control, IEEE Transactions on*, 46(4):1001–1012, 1999. 119

[91] A. Leonard, H. Krag, G. Lachapelle, K. O'Keefe, C. Huth, and C. Seynat. Impact of GPS and Galileo Orbital Plane Drifts on Interoperability Performance Parameters. In *European Navigation Conference*, pages 22–25, Graz, Austria, 2003. 138

[92] F. Levi, A. Godone, S. Micalizio, and C. Calosso. CPT Maser clock evaluation for Galileo. In *34th Annual Precise Time and Time Interval (PTTI) Meeting*, pages 139–150, Reston,Virginia, USA, December 2002. 59

[93] M. W. Levine. Performance of the GPS Cesium beam frequency standards in orbit. In *35th Frequency Control Symposium*, pages 651–656, Beverly, Massachusetts 01915, USA, May 1981. 53

[94] W. Lewandowski and Z. Jiang. Use of GLONASS at the BIPM. In *41st Annual Precise Time and Time Interval (PTTI) Meeting*, pages 5–14, Santa Ana Pueblo, New Mexico, USA, November 2009. 17

[95] R. Lutwak, D. Emmons, R. Garvey, and P. Vlitas. Optically-pumped cesium beam frequency standard for GPS-III. In *33rd Annual Precise Time and Time Interval (PTTI) Meeting*, pages 19–32, Long Beach, California,USA, November 2001. 55

[96] M. Malik, G. Gatti, Valter, R. Kieffer, and G. Robertson. GIOVE-B satellite and payload overview. In *European Navigation Conference*, pages 22–29, Toulouse,France, April 2008. 66

[97] L. A. Mallette, P. Rochat, and J. White. Historical review of atomic frequency standards used in space systems 10 year update. In *38th Annual Precise Time and Time Interval (PTTI) Meeting*, Los Angeles, CA 90009, USA, January 2007. 53, 58

[98] L. A. Mallette, P. Rochat, and J. White. Space qualified frequency sources (clocks) for current and future GNSS applications. In *41st Annual Precise Time and Time Interval (PTTI) Meeting*, Reston, Virginia, USA, January 2010. 54, 57, 58, 61

[99] H. S. Margolis. Optical frequency standards and clocks. *Contemporary Physics*, 51(1):37–58, 2010. 62

[100] W. Markowitz, R. G. Hall, L. Essen, and J. V. L. Parry. Frequency of cesium in terms of ephemeris time. *Phys. Rev. Lett.*, 1(3):105–107, Aug 1958. 14

[101] D. Matasakis, M. Miranian, and P. Koppang. New steering strategies for the USNO master clocks. In *31st Annual Precise Time and Time Interval (PTTI)*, pages 277–284, Dana Point, California, USA, December 2000. 65

[102] E. Mattison and R. Vessot. Design of a hydrogen maser for space. In *12th Ann. Precise Time and Time Interval Meeting*, pages 463–470, Reston, VA 20190,USA, 1996. 60

[103] D. McCarthy. CCTF/09-32, note on coordinated universal time. Technical report, BIPM, 2009. 15

[104] D. D. McCarthy and G. Petit. IERS conventions 2003, Chapter 10:General relativistic models for space-time coordinates and equations of motion. Technical Report 32, IERS, 2003. 36, 42, 46

[105] T. B. McCaskill, J. A. Buisson, M. M.Largay, and W. G.Reid. On-orbit frequency stability analysis of the GPS NAVSTAR-1 quartz clock and the NAVSTARs-6 and -8 rubidium clocks. In *35th Frequency Control Symposium*, pages 103–125, NASA Goddard Space Flight Center Greenbelt, MD,USA, November 1984. 62

[106] J. H. Meeus. *Astronomical Algorithms*. Willmann-Bell, Incorporated, 1999. 138

[107] T. Melgard, R. Zandbergen, J. F. Sanchez, and L. Agrotis. First real time GPS/GLONASS orbit/clock decimetre level precise positioning service. In *European Navigation Conference*, pages 1–5, Naples, Italy, May 2009. 173

[108] M. J. V. Melle. Cesium and rubidium frequency standards status and performance on the GPS program. In *27th Annual Precise Time and Time Interval (PTTI) Meeting*, pages 167–180, Falcon Air Force Base Colorado Springs, CO, USA, November 1995. 54, 57

[109] S. Micalizio. Pulsed optically pumped Rb clock with optical detection: first results. In *24th European Frequency and Time Forum (EFTF)*, (ESA/ESTEC) Noordwijk, The Netherlands, April 2010. 59

[110] G. Mileti and P. Thomann. Study of the S/N performance of passive atomic clocks using a laser-pumped vapour. In *European Frequency and Time Forum (EFTF)*, pages 271–276, Besançon, France, January 1995. 58

[111] O. Montenbruck, A. Hauschild, P. Steigenberger, and R. B. Langley. Three's the challenge. *GPS World*, July 2010. 91, 146

[112] O. Montenbruck, P. Steigenberger, E. Schönemann, A. Hauschild, U. Hugentobler, and R. Dach. Flight characterization of new generation GNSS satellite clocks. In *International Technical Meeting of the Satellite Division of the Institute of Navigation (ION GNSS)*, Portland Oregon, USA, September 2011. IEEE. iii, 184

[113] B. Moreno. *Development of Algorithms for the GNSS data processing: their application to the Modernized GPS and Galileo Scenarios*. Diploma thesis, Universidad Complutense de Madrid, November 2011. 122

[114] A. Mozo, R. Piriz, V. Fernandez, D. Navarro, and F. Gonzalez. Giove orbit and clock determination and prediction: Experimentation results. In *European Navigation Conference*, Toulouse,France, April 2008. 8

[115] Navstar GPS Joint Program Office. *Navstar GPS Space Segment. Navigation User Interfaces, IS-GPS-200, Revision D*, December 2004. 16, 18, 20, 93

[116] A. A. Nielsen. Least squares adjustment: Linear and nonlinear weighted regression analysis. *Danish National Space Center/Informatics and mathematical modelling, Technical Univ. of Denmark*, 2012. 79

[117] Novatel. *Galileo Test Receiver (GTR)*, April 2010. 8, 115

[118] Novatel. *GNSS wideband antenna :GNSS-750*, February 2010. 8

[119] J. Oaks, M. Largay, W. Reid, and J. Buisson. Comparative analysis of GPS clock performance using both code-phase and carrier-derived pseudorange observations. In *36th Precise Time and Time Interval (PTTI) Meeting*, pages 431–440, Washington. DC 20001,USA, December 2004. 119

[120] J. Oaks and M. M. Largay. Global positioning system constellation clock performance. In *34th Annual Precise Time and Time Interval (PTTI) Meeting*, pages 77–88, Reston, Virginia, USA, December 2004. 49, 94

[121] V. Oehler, F. Luongo, H. Trautenberg, J. Boyero, J. Krueger, and T. Rang. The Galileo integrity concept and performance. In *European Navigation Conference*, pages 19–22, Munich, Germany, July 2005. 180

[122] D. Orgiazzi, P.Tavella, and F. Lahaye. Experimental assessment of the time transfer capability of precise point positioning. In *37th Precise Time and Time Interval (PTTI) Meeting*, Vancouver, Canada, August 2005. 105

[123] B. Parkinson and S. T. Powers. The origins of GPS, part 2: fighting to survive. *GPS World*, June 2010. 56

[124] B. Parkinson, J. Spilker, P.Axelrad, and P.Enge. *Global positioning system: theory and applications*. Progress in astronautics and aeronautics, March 1996. 43, 52, 53, 63

[125] M. Pearlman, J. Degnan, , and J. Bosworth. The international laser ranging service. *Advances in Space Research*, 30(2):135–143, 2002. 110

[126] J. Perello, N. Batzilis, G. Lopez, and J. Alegre. GNSS payload and signal characterization using a 3m dish antenna. In *International Technical Meeting of the Satellite Division of the Institute of Navigation (ION GNSS)*, Nashville, Tenessee, USA, September 2012. IEEE. 68

[127] PerkinElmer Optoelectronics. *Data Sheet RFS*, April 2010. 57

[128] G. Petit. Report of the bipm/iau joint committee on relativity for space-time reference systems and metrology. Technical report, DTIC Document, 2000. 16

[129] G. Petit. Values and uncertainties of the hardware delays of BIPM geodetic systems and estimation of the type B uncertainty of P3/PPP link calibrations. Technical report, BIPM, December 2009. ftp://tai.bipm.org/TFG/CALIB_GEO/tm172.pdf. 115

[130] G. Petit and E. F. Arias. Use of IGS products in TAI applications. *Journal of Geodesy*, 83(3-4), July 2009. 77

[131] J. Phelan, T. Dass, G. Freed, J. Rajan, J. D'Agostino, and M. Epstein. GPS block IIR clocks in space: current performance and plans for the future. In *Frequency Control Symposium and Exposition. Proceedings of the 2005 IEEE International*, page 19. IEEE, 2005. 25

[132] D. Piester, A. Bauch, J. Becker, and A. Hoppmann. Time and frequency broadcast with dcf77. In *43rd Annual Time and Time Interval (PTTI) Systems and Applications Meeting*, pages 185 – 196, Long Beach, California, USA, November 2012. 73

[133] D. Piester, M. Fujieda, M. Rost, and A. Bauch. Time transfer through optical fibers (TTTOF): first results of calibrated clock comparisons. In *41st Annual Precise Time and Time Interval (PTTI) Meeting*, Santa Ana Pueblo, New Mexico, USA, November 2009. 77

[134] R. Piriz, D. Calle, A. Mozo, P. Navarro, D. Rodríguez, and G. Tobías. Orbits and clocks for GLONASS precise-point-positioning. In *International Technical Meeting of the Satellite Division of the Institute of Navigation (ION GNSS)*, pages 1–10, Georgia, USA, May 2009. 173

[135] R. Piriz, M. Cueto, and V. Fernandez. GPS/Galileo interoperability: GGTO, timing biases, and Giove-A experience. In *38th Annual Precise Time and Time Interval (PTTI) Meeting*, pages 49–68, Washington. DC 20001,USA, Dec 2006. 111

[136] R. Piriz, V. Fernandez, P. Tavella, I. Sesia, G. Cerretto, M. Falcone, D. Navarro, J. Hahn, F. Gonzalez, M. Tossaint, and M. Gandara. The Galileo system test bed V2 for orbit and clock Modeling. In *International Technical Meeting of the Satellite Division of the Institute of Navigation (ION GNSS)*, pages 549–562, Fort Worth Convention Center, Fort Worth, Texas, 2006. 5, 150, 151

[137] J. Prestage and L. Maleki. Space flyable Hg+ frequency standards. In *48th.Frequency Control Symposium, Proceedings of the 1994 IEEE International*, pages 747–754. IEEE, 1994. 61

[138] H. Rawicz, M. Epstein, and J. Rajan. The time keeping system for GPS Block IIR. In *24th Precise Time and Time Interval (PTTI) Meeting*, pages 5–16, McLean, Virginia,USA, 1992. 63, 142

[139] J. Ray and J. Griffiths. Status of IGS ultra-rapid products for real-time applications. In *International GNSS Service Analysis Center Workshop*, Miami Beach, USA, June 2008. 173

[140] J. Ray and K. Senior. Geodetic techniques for time and frequency comparisons using GPS phase and code measurements. *Metrologia*, 42(4):215, 2005. 103

[141] V. S. Reinhardt and C. B. Sheckells. Redundant atomic frequency standard time keeping system with seamless AFS switchover. In *31st Precise Time and Time Interval (PTTI) Meeting*, pages 101–110, Dana Point, California,USA, 1999. 65

[142] S. Revnivykh. GLONASS Status and Progress. In *International Technical Meeting of the Satellite Division of the Institute of Navigation (ION GNSS)*, September 2010. 54

[143] J. Riedesel. GPS operations civil GPS service interface committee (CGSIC) brief. Technical report, 2d Space Operations Squadron, September 2008. 57

[144] W. J. Riley. A rubidium clock for GPS. In *13th Precise Time and Time Interval (PTTI) Meeting*, pages 609–630, Naval Research Laboratory, Washington, USA, Dec 1981. 57

[145] W. J. Riley. Rubidium atomic frequency standards for GPS Block IIR. In *24th Precise Time and Time Interval (PTTI) Meeting*, pages 221–230, McLean, Virginia, USA, Sept 1992. 57

[146] P. Rochat, F. Droz, P. Mosset, G. Barmaverain, Q. Wang, and D. Boving. The onboard Galileo rubidium and passive maser, status and performance. In *20th European Frequency and Time Forum*, Besançon, France, March 2005. 58, 59

[147] E. Rooney, M. Unwin, A. Bradford, and P. Davies. Meet GIOVE-A Galileo first test satellite. *GPS World*, May 2007. 66

[148] Russian Institute of Space Device Engineering. *GLONASS Interface Control Document*, 5.1 edition, 2008. 16

[149] K. Senior, P. Koppang, and J. Ray. Developing an IGS time scale. *Ultrasonics, Ferroelectrics and Frequency Control, IEEE Transactions on*, 50(6):585–593, 2003. 85

[150] K. Senior, J. R. Ray, and R. L. Beard. Characterization of periodic variations in the GPS satellite clocks. *GPS Solutions*, 12(3):221–225, July 2008. iii, 135, 144, 150, 152, 185

[151] Septentrio. *GeNeRx1 GPS Galileo receiver datasheet*. Septentrio, June 2006. 8, 115

[152] A. Simsky. Three's the charm: triple-frequency combinations in future GNSS. *Inside GNSS*, pages 38–41, July 2006. 112

[153] A. Simsky, D. Mertens, J.-M. Sleewaegen, W. D. Wilde, M. Hollreiser, and M. Crisci. Multipath and tracking performance of Galileo ranging signals transmitted by GIOVE-B. In *International Technical Meeting of the Satellite Division of the Institute of Navigation (ION GNSS)*, Savannah, Georgia,USA, September 2008. 112, 114

[154] Spectra Time. *iSource+ Space-Qualified RAFS Specifications*, March 2004. 141, 142

[155] Spectra Time. *iMaser 3000, smart active hydrogen maser clock data sheet*, 2011. www.spectratime.com/documents/iMaser_Clock_Spec.pdf. 103, 104, 105

[156] X. Stehlin, Q. Wang, F. Jeanneret, and P. Rochat. Galileo system time physical realization. In *38th Annual Precise Time and Time Interval (PTTI) Meeting*, Washington. DC 20001,USA, 2006. 16, 65

[157] P. Steigenberger, U. Hugentobler, O. Montenbruck, and A. Hauschild. Precise orbit determination of GIOVE-B based on the CONGO network. *Journal of Geodesy*, 85:357–365, 2011. 10.1007/s00190-011-0443-5. 107, 110

[158] E. Swift and B. Hermann. Orbit period frequency variations in the GPS satellite clocks. In *10th Ann. Precise Time and Time Interval (PTTI) Meeting*, pages 87–100, Vienna, Virginia, USA, December 1988. iii, 135, 152, 185

[159] Symmetricom. *CAFS 4415 Specifications*, 2010. http://www.gigatest.net/symmetricom TT&M/Space, Defense & Avionics/ds_4415.pdf. 54

[160] Symmetricom. *Chip scale atomic clock (CSAC) data sheet - SA.45s*, 2010. 33

[161] Symmetricom. *MHM 2010, Active Hydrogen Maser data sheet*, 2011. 104, 105

[162] A. Tetewsky. Making sense of inter-signal corrections. *Inside GNSS*, pages 37–48, August 2009. 65, 93, 115

[163] Thales Alenia Space Spain. *Clock Monitoring and Control Unit(CMCU) specifications*, April 2011. http://www.thalesgroup.com/Countries/Spain/What_we_do/-Space/Telecommunication_and_Navigation. 64

[164] D. Thaller, A. Steinbach, and R. Dach. Combined analysis of GNSS and SLR observations for the GIOVE satellites. In *EGU General Assembly*, May 2009. 110

[165] S. Thoelert. In-Orbit analysis of antenna pattern anomalies of GNSS. In *International Technical Meeting of the Satellite Division of the Institute of Navigation (ION GNSS)*, Portland, Oregon, USA, September 2010. 113

[166] B. Titus, J. Betz, C. Hegarty, and R. Owen. Intersystem and intrasystem interference analysis methodology. *Proceedings of IONGNSS-2003*, 2003. 98

[167] G. Tobias, I. Hidalgo, D. Rodríguez, Álvaro Mozo, S. Binda, F. González, A. Mudrak, P. Tavella, I. Sesia, and G. Cerretto. Building Galileo navigation system: two years of GIOVE-M experimentation. In *International Technical Meeting of the Satellite Division of the Institute of Navigation (ION GPS)*, Savannah, Georgia, USA, September 2009. 99, 102

[168] USA. GNSS evolutionary architecture study. Technical report, USA Federal Aviation Administration (FAA), February 2008. 170

[169] W. Vodel, H. Koch, S. Nietzsche, J. Zameck Glyscinski, R. Neubert, and H. Dittus. Testing einstein's equivalence principle at bremen drop tower using lts squid technique. *Applied Superconductivity, IEEE Transactions on*, 11(1):1379 –1382, mar 2001. 39

[170] P. Waller, F. Gonzalez, S. Binda, I. Sesia, I. Hidalgo, G. Tobias, and P. Tavella. The in-orbit performances of giove clocks. *IEEE Transactions on Ultrasonics, Ferroelectrics, and Frequency Control*, 57(3), March 2010. 107

[171] J. White and R. Beard. Space clocks - why they're different. In *33rd Annual Precise Time and Time Interval (PTTI) Meeting*, pages 7–18, Long Beach, California,USA, November 2001. 51

[172] J. White, A. Frank, and V. Folen. Passive maser development at NRL. In *12th Ann. Precise Time and Time Interval Appl. and Planning Meeting (PTTI)*, volume 1, pages 495–513, Washington, D.C.,USA, 1981. 60

[173] W. D. Wilde, F. Boom, J.-M. Sleewaegen, and F. Wilms. Tracking china's meo satellite on a hardware receiver. *GPS World*, July 2007. Septentrio. 8

[174] W. D. Wilde, J.-M. Sleewaegen, K. V. Wassenhove, and F. Wilms. A First-of-a-Kind Galileo receiver breadboard to demonstrate Galileo tracking algorithms and performances. In *International Technical Meeting of the Satellite Division of the Institute of Navigation (ION-GNSS)*, Long Beach, California, USA, September 2004. 8

[175] M. S. Willard Marchis. GPS-III bringing new capabilities to the global community. *Inside GNSS*, pages 34–48, September 2011. 65

[176] B. D. Wilson, C. H. Yinger, W. A. Feess, and C. Shank. New and improved: The broadcast Interfrequency biases. *GPS world*, September 1999. 65

[177] P. Wolf and G. Petit. Relativistic theory for clock syntonization and the realization of geocentric coordinate times. *Astronomy and Astrophysics*, 304:653–661, Dec. 1995. 40

[178] P. R. Wolf and C. D. Ghilani. *Adjustment computations: Statistics and least squares in surveying and GIS*. John Wiley & Sons New York, 1997. 79

[179] A. Wu. Performance evaluation of the GPS block IIR time keeping system. In *28th Precise Time and Time Interval (PTTI) Meeting*, pages 441–454, Reston, Virginia,USA, December 1996. 63, 142

[180] A. Wu. Investigation of the GPS Block IIR Time Keeping System (TKS) Anomalies Caused by the Voltage-Controlled Crystal Oscillator (VCXO). In *31st Precise Time and Time Interval (PTTI) Meeting*, pages 55–64, Dana Point, California,USA, 1999. 63, 64, 142

[181] A. Wu and Feess. Development and evaluation of GPS space clocks for GPS III and beyond. In *32nd Precise Time and Time Interval (PTTI) Meeting*, pages 389–400, Reston, Virginia,USA, November 2000. 54, 142

[182] W. Zhang, V. Venkatasubramanian, H. Liu, M. Phatak, and S. Hahn. SirF instant fix II technology. In *International Technical Meeting of the Satellite Division of the Institute of Navigation (ION GNSS)*, Savananah,Geogia,USA, September 2008. 173